# RECENT WORKS ON MICROBES AND INFECTIONS IN CHINA

## Selected from the Journal of Microbes and Infections (China)

# RECENT WORKS ON MICROBES AND INFECTIONS IN CHINA

## Selected from the Journal of Microbes and Infections (China)

edited by

Yu-Mei Wen
Fudan University, China

Shan Lu
University of Massachusetts Worcester, USA

Yi-Wei Tang
Vanderbilt University, USA

Xin-Hua Weng
Fudan University, China

World Scientific

NEW JERSEY • LONDON • SINGAPORE • BEIJING • SHANGHAI • HONG KONG • TAIPEI • CHENNAI

*Published by*

World Scientific Publishing Co. Pte. Ltd.
5 Toh Tuck Link, Singapore 596224
*USA office:* 27 Warren Street, Suite 401-402, Hackensack, NJ 07601
*UK office:* 57 Shelton Street, Covent Garden, London WC2H 9HE

**British Library Cataloguing-in-Publication Data**
A catalogue record for this book is available from the British Library.

**RECENT WORKS ON MICROBES AND INFECTIONS IN CHINA**
**Selected from the Journal of Microbes and Infections (China)**



ISBN-13 978-981-283-566-6
ISBN-10 981-283-566-0

Typeset by Stallion Press
Email: enquiries@stallionpress.com

Printed in Singapore by Mainland Press Pte Ltd.

# Preface I

Infectious diseases are of global concern. They have been such in the past and will continue to be such in the future. Multidrug-resistant tuberculosis, malaria, AIDS, viral hepatitis, pandemic influenza and other emerging infections are just a few of the problems the world is facing. For China, the country with the largest population in the world, most of these diseases are a particular challenge. A dramatic example is the SARS outbreak, which has told us that, in order to successfully control infections, increased efforts are necessary to understand the disease mechanisms and epidemiology and to develop new tools for diagnosis, prophylaxis and therapy. This volume is an impressive document on the enormous progress that Chinese research has made in the last couple of years in all of these fields. Since, so far, only some of these studies have been published in international journals, a lot of information has not spread beyond the borders of the country. I therefore warmly welcome the opportunity to introduce this volume, which gives a broader overview to the outside world of what Chinese bacteriologists and virologists are doing.

*Hans Dieter Klenk*
*Universität Marburg, Germany*
*October 28, 2008*

# Preface 2

Thirty years ago, when China resumed its normal environment for scientific activities, communication with the outside world began. A number of eminent virologists, bacteriologists and molecular biologists, including the late Joseph Melnick, Pierre Tiollais, Morris Pollard and my many other friends, visited China, and I had the opportunity to be their scientific interpreter. Besides, young Chinese students and scientists were sent abroad to study, which was hugely beneficial for China in proceeding at a rapid pace during the 30 years. In 1982, my mentor, the late Prof. Fei-Qin Lin, started to edit a journal named *Medical Microbiology Abroad*, which published reviews and abstracts introducing new progress in medical microbiology from journals published in various languages, including English, French, Russian and Japanese. The editorial board of *Medical Microbiology Abroad* consisted of medical microbiologists from many provinces of China, and the journal was well received. *Medical Microbiology Abroad* has been an important reference journal providing scientific information on medical microbiology for scientists and mostly graduate students throughout China, especially for those in regions where foreign journals were not easily accessible.

After more than 20 years of publishing *Medical Microbiology Abroad*, the editorial board realized that since the Internet is widely used, people can read journals on the websites. At the same time, more and more research on medical microbiology is being conducted, and to promote translational research we need to set up close contact among microbiologists, clinicians, veterinarians, epidemiologists, vaccinologists and health workers. Therefore, we decided to change *Medical Microbiology Abroad* into the *Journal of Microbes and Infections*, a journal publishing research articles, reviews and scientific news on microbes and infections in humans and animals. The current editorial board has been expanded to include microbiologists, immunologists, epidemiologists, medical doctors and veterinarians. The first issue of the *Journal of Microbes and Infections* (in Chinese) was published in 2006. After close to three years of hard work, we have been looking for opportunities to introduce part of the work in China to the outside world. Now, our effort is supported by World Scientific Publishing Company, which decided to publish selected articles from the journal in English. We consider this a good approach to building up communication between Chinese scientists and scientists abroad. We hope that this window will open up a supplementary track for scientists outside of China to know in detail the work on microbes and infections carried out in China, and perhaps this journal can also serve as a matchmaker for future collaboration among scientists. Due to the limitation of space, the material and methods are omitted from this book. However, readers can contact the authors for details.

*Yu-Mei Wen*
*Key Laboratory of Medical Molecular Virology*
*Institute of Medical Microbiology*
*Institutes of Biomedical Sciences*
*Shanghai Medical College*
*Fudan University, Shanghai, China*

# Contents

# SECTION 1

# Bacteria

# Introduction and Comments: Studies on Bacteria by Molecular and Experimental Approaches

Yu-Mei Wen

*Key Laboratory of Medical Molecular Virology,*
*Shanghai Medical College, Institutes of Biomedical Sciences,*
*Fudan University, Shanghai 200031, China*

When the first paper on the full-length genomic sequence of *Hemophilus influenza* was published in 1995, Chinese bacteriologists, molecular biologists and scientists working in bioinformatics collaborated and started to study bacteria on a genomic basis. To date, there are around 15 strains of medically associated bacteria being sequenced by Chinese scientists alone or in collaboration with scientists abroad, and the bacterial strains being sequenced include *Leptospira interrogans*, *Shigella flexneri*, *Staphylococcus epidermidis*, *Streptococcus suis* type 2 and *Bacillus anthrax*. Genome-based research on bacteria opened a new era of basic bacterial studies, which may lead to extraordinary changes in future microbial pathogenesis diagnosis, prevention and treatment. This has also influenced the trend of basic research on bacteria in China. In this volume, based on the full-length sequencing of *Leptospira interrogans*, two papers (Chapter 1.1, Chapter 1.2) reflect the two trends of research on this microbe: one reflects the search for novel proteins of this microorganism which may be employed to improve the future vaccine or as biomarkers for the disease, while the other paper reports an epidemiological survey of this microbe in murine reservoirs, which is important for analyzing and predicting the epidemiology of the disease. Another direction in the genomic study of bacteria was to screen for inhibitors of the virulent gene or enzymes of bacteria. Chen *et al.* (Chapter 1.5) in this volume describe the effects of inhibitors of the EnvZ histidine kinase of *Shigella flexneri*, showing that screening for new drug targets is one of the major fields for research on bacteria.

As the number of tuberculosis cases is increasing, aside from developing rapid and sensitive diagnostic methods, basic research aiming at developing new drug targets and new candidates for vaccine are interests shared by scientists at home and abroad. The papers by Jin *et al.* (Chapter 1.3) describe the cloning and purification of a dehydrogenase from *Mycobacterium tuberculosis*. However, these studies are relatively preliminary. Functional studies of this enzyme should be further explored, and effects of blocking bacterial enzymes or the immunogenicity of bacterial enzymes are needed to understand the roles of this enzyme in *M. tuberculosis*. Sun's paper (Chapter 1.4) on *Streptococcus suis* predicts the *fbps* gene as an adhesin without an anchoring component. Similarly, functional studies and implications of this finding are necessary.

In addition to molecular and experimental studies, practical application of genetic study has been conducted. Food safety is of the utmost importance, as food contaminated by microbes will cause fear and even turmoil in the community. Yang *et al.* (Chapter 1.6) reported studies on the nonculturable form of *Campylobacter jejuni* by storage at 0°C, frozen at –70°C, and starvation and a cytotoxic distending toxin (CDT) virulence assay could be used to detect VBNC *C. jejuni*; they found that though the virulence was lower in VBNC *C. jejuni* compared to that of the bacterial state, VBNC *C. jejuni* was still toxic to cells. This attempt to study bacterial contamination of food at low temperature shows that scientists are concerned about food safety and should be encouraged. As data on the survey of drug resistance among bacteria in China have been reported elsewhere, they are not included in this volume.

It was once thought that bacteriology had come to an end, and should be substituted by molecular biology. In contrast to what was thought 10–15 years ago, current studies on bacteria are well supported by grants in China. Both molecular research and application research on bacteria are encouraged. We expect that research on emerging microbes or microbial variants will be reported in the future from China, and that functional genomic research on bacteria will be conducted in depth. At the same time, translational research will be more widely pursued to improve human health.

# 1.1

# Cloning, Expression and Immunological Evaluation of a Novel Protein Belonging to the OmpA Family in *Leptospira interrogans*

Wei Lin[*,†,§], Hong-Liang Yang[*,‡], Yang Yang[*], Bao-Yu Hu[*], Li-Zhi Tan[†] and Xiao-Kui Guo[*,¶]

[*]*Department of Medical Microbiology and Parasitology, Institute of Medical Sciences, Shanghai Jiao Tong University, School of Medicine, Shanghai 200025, China*
[†]*Institute of Pathogenic Biology, Nan Hua University, Hengyang, Hengyang 421001, China*
[‡]*Institute of Plant Physiology and Ecology, Shanghai Institutes for Biological Sciences, Chinese Academy of Sciences, Shanghai 200032, China*
[§]*Licheng Center for Disease Control and Prevention, Jinan 250100, China*

**Objective.** To clone and express a novel OmpA gene (LA0301) of the *Leptospira interrogans* serogroup Icterohaemorrhagiae serovar Lai strain Lai, and to analyze the antigenicity and conservation of the recombinant protein in 15 *L. interrogans* strains belonging to 15 prevalent serogroups of *L. interrogans* in China.

**Methods.** The characteristics of LA0301 were predicted using bioinformatics software. The prokaryotic expression system of LA0301 was

[¶]Corresponding author. Email: microbiology@sjtu.edu.cn

constructed by routine molecular-biological techniques, and expression of the recombinant protein was identified by SDS-PAGE and Western blot after induction with IPTG. Western blot was used to identify the antigenicity and the conservation of expressed LA0301 in 15 *L. interrogans* strains belonging to 15 prevalent serogroups in China.

**Results.** Bioinformatics predicted that LA0301 is a novel outer membrane protein belonging to the OmpA family in *L. interrogans*. The prokaryotic expression system of LA0301 was constructed successfully. The titer of antirecombinant LA0301 immunized BALB/c mouse sera was 1:32000. The LA0301 protein was detected in all 15 strains of *L. interrogans*. The antibody versus LA0301 was also detected by Western blot in rabbit sera immunized with *L. interrogans* strain Lai.

**Conclusion.** LA0301, a novel protein belonging to the OmpA family in *L. interrogans*, not only showed good antigenicity and conservation among strains from different serotypes, but also induced antibody responses in rabbits during infection with *L. interrogans* strain Lai. This study provides information on a vaccine candidate against leptospirosis.

*Keywords*: *Leptospira interrogans*; OmpA; vaccine candidate gene.

## 1. Introduction

China is one of the countries where leptospirosis is most prevalent. However, the current commercially available vaccines against this disease need improvement; for example, they cannot provide cross-protection against infections with a different *Leptospira* serovar, and the vaccinees need to be boosted every year in areas of high prevalence. Therefore, it is necessary to develop a more effective vaccine against this worldwide zoonosis. Some of the outer membrane proteins (OMPs) of *Leptospira* are not only toxins, but also specific antigens which can elicit specific immune responses.[1] LipL32 (Hap1) and LipL41 have been demonstrated to be vaccine candidates, but satisfactory effects have not been achieved.[2,3]

The proteins of the OmpA family play important roles in invasion, adhesion, and maintenance of the integrity of the outer membrane.[4] The C terminus contains a peptidoglycan-associated domain, which appears to connect the outer membrane and the peptidoglycan. There are eight OMPs with OmpA-associated domains in the *Leptospira interrogans* serogroup Icterohaemorrhagiae serovar Lai strain Lai. For example, Loa22[5] and OmpA52[6] are both OmpA family proteins and have good antigenicity. Yang *et al.* have screened 226 candidate genes from the genome of *L. interrogans* serovar Lai strain Lai by combination techniques of bioinformatics, comparative genomics and transcriptions. OmpA, coded by the LA0301 gene, was one of the 120 vaccine candidates that were predicted to be located in the outer membrane.[7] In this study, LA0301 was expressed in *E. coli*, and the immunoreactivity (antigenicity) and immunogenicity of the recombinant LA0301 protein were studied. In addition, conservatism of the recombinant protein was analyzed among 15 *L. interrogans* strains belonging to 15 prevalent serogroups in China. Moreover, the antibody versus LA0301 was detected in rabbit sera of anti-*L. interrogans* strain Lai by Western blot.

## 2. Results

### 2.1. *Bioinformatics analysis*

The full length of the LA0301 gene is 1770 bp, coding 589 amino acid residues. The *in silico* analysis of the LA0301 protein showed that there was no α-helix transmembrane topology, but that a signal peptidase I site existed at the 21 and 22 amino acid residue. Moreover, an OmpA family domain (pfam: 00619) from 504 to 580 amino acid residues was predicted, which is highly homologous with OmpA family protein of *M. tuberculosis*.

### 2.2. *Cloning and expression of the recombinant LA0301 in* E. coli

The recombinant LA0301 protein was successfully expressed in *E. coli* M15 cells by using plasmid pQE31-LA0301, which contains the

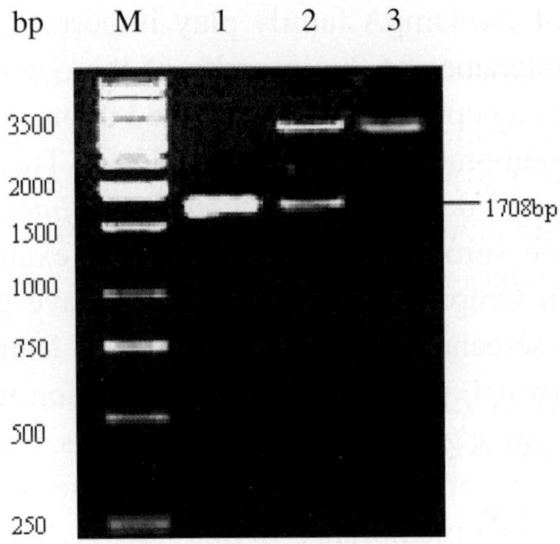


**Fig. 1.** Restriction enzyme analysis of recombinant plasmid with *Bam*HI and *Pst*I. (1) PCR production, (2) recombinant plasmid digested with *Bam*HI and *Pst*I, (3) empty vector PMZ, (M) DNA marker.

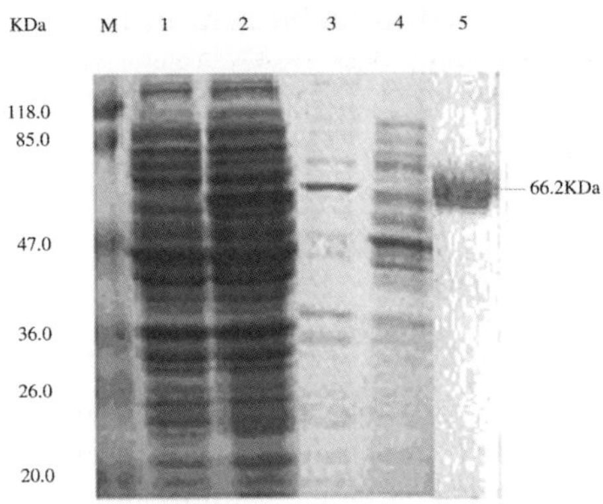


**Fig. 2.** Expression, purification and identification of recombinant LA0301 in *E. coli* M15. Numbers at left indicate molecular mass in kilodaltons. (1) Uninduced *E. coli* M15, (2) IPTG-induced *E. coli* M15, (3) precipitation of recombinant protein, (4) supernatant of recombinant protein, (5) purified protein, (M) protein ladder.

LA0301 gene (Fig. 1). SDS-PAGE and Western blot were used to identify the recombinant protein. Western blot analysis results showed that the molecular weight of this protein is approximately 66 kDa (Fig. 2).

## 2.3. *The titer of serum anti-LA0301 and immunoreactivity of the recombinant LA0301*

Humoral immune response to these proteins was shown by ELISA assay of antibodies in anti-LA0301 mouse sera. The antibody titer was 1:32000 by ELISA. The immunoreactivity of the recombinant LA0301 was also detected by Western blot analysis (Fig. 3A).

## 2.4. *The antibody of LA0301 was detected in anti-*L. interrogans *rabbit sera*

The antibody of LA0301 was detected in anti-*L. interrogans* rabbit sera using Western blot analysis (Fig. 3B).

## 2.5. *Conservatism analysis of LA0301 in 15 different* L. interrogans *strains*

Western blot results showed that there were specific LA0301 bands in all tested *L. interrogans* strains belonging to the 15 prevalent serogroups in China. Nevertheless, there were slight differences in molecular weight among the 15 prevalent *L. interrogans* serogroups (Fig. 4).

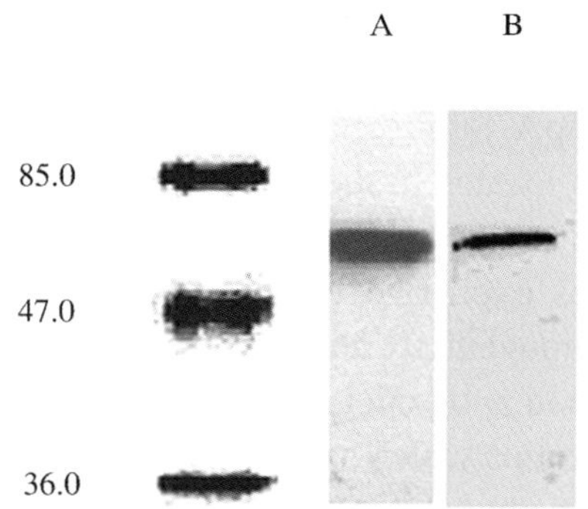


**Fig. 3.** Western blot of anti-LA0301 mouse sera with LA0301 protein (A) and anti-*L. interrogans* strain Lai rabbit sera with LA0301 protein (B).

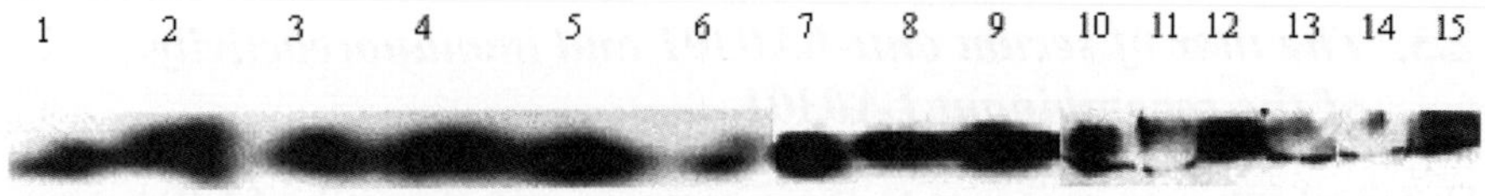


**Fig. 4.** Detection of LA0301 in 15 prevalent *Leptospira* strains using Western blot. (1) 56601, (2) 56602, (3) 56603, (4) 56604, (5) 56605, (6) 56606, (7) 56607, (8) 56608, (9) 56609, (10) 56610, (11) 56612, (12) 56613, (13) 56615, (14) 56635, (15) 56655. The numbers are *L. interrogans* strain names.

## 3. Discussion

OmpA is a highly conserved OMP family in gram-negative bacteria, which not only induces specific humoral and cellular immunity, but also can stimulate dendritic cell to secrete cytokines.[8,9] Moreover, it is involved in the assistance of antigen presentation, cross-presentation and cell adhesion.[10] There are eight genes encoding the OmpA-associated domain in the *L. interrogans* serovar Lai strain Lai genome, i.e. LA0056, LA0222, LA0301, LA2215, LA3615, LA3685, LA4337 and LB328. Koizumi *et al.* screened the LA0222 (Loa22) gene from *L. manilae* UP-MMC via an alkaline phosphatase promoter A fusion expression system and immunohybridization. The LA0222 encoding protein was documented as a novel virulence factor of pathogenic *Leptospira* by immunoblotting.[5] Ristow *et al.* verified that Loa22 played an important role in *Leptospira* infection by building in a HimarI transposon to induce Loa22 random mutation.[11] Omp52 (LA3615) was identified as a good immunogenic antigen, as it is located in the outer membrane.[6] In addition, Yang *et al.* identified that there were four OmpA family vaccine candidates (LA0222, LA0301, LA3615 and LB328) located in the outer membrane.[7]

In our study, the LA0301 protein, one of many *L. interrogans* OmpA family members, was expressed by using the plasmid pQE31-LA0301. The analysis of immunogenicity and conservatism showed that the recombinant protein has good antigenicity and conservation among 15 different *L. interrogans* serogroups which are prevalent in China. These results are consistent with those obtained in a comparative genomic hybridization study.[7] In addition, the LA0301 antibody was also detected in rabbit anti-*L. interrogans* sera using Western blot analysis, indicating that the LA0301 protein is most likely an OMP, which may promote the production of the antibody when the host is infected by *Leptospira*. In conclusion, the current study on LA0301 provides a basis for the development of novel vaccine candidates against leptospirosis.

## References


1. Vachaev BF *et al.*, *Zh Mikrobiol Epidemiol Immunobiol*, 1992, **11–12**: 46–49.
2. Ruan P *et al.*, *Chin J Microbiol Immunol*, 2005, **25**: 369–374.
3. Haake DA *et al.*, *Infec Immun*, 1999, **67**: 6572–6582.
4. Jeannin P *et al.*, *Nat Immunol*, 2000, **1**: 502–509.
5. Koizumi N *et al.*, *FEMS Microbiol Lett*, 2003, **226**: 215–219.
6. Hsieh WJ *et al.*, *FEMS Microbiol Lett*, 2005, **243**: 339–345.
7. Yang HL *et al.*, *BMC Genomics*, 2006, **7**: 1–12.
8. Dabo SM *et al.*, *Microb Pathog*, 2003, **35**: 147–157.
9. Torres AG *et al.*, *Infec Immun*, 2006, **74**: 2676–2685.
10. Jeannin P *et al.*, *Immunity*, 2005, **22**: 551–560.
11. Ristow P *et al.*, *PLoS Pathog*, 2007, **13**(3): 894–904.

# 1.2

# Surveillance of Leptospirosis and the Murine Reservoirs in China, 2005

Xiu-Gao Jiang[*,§], Xiang-Yan Zhang[†,§], Wen-Jun Zhong[‡], Yi-Xin Nie[*], Wen-Wu Yin[‡], Xiu-Wen Li[*] and Xiao-Kui Guo[†,¶]

[*]*National Institute for Communicable Disease Control and Prevention, Chinese Center for Disease Control and Prevention, Beijing 102206, China*
[†]*Department of Medical Microbiology and Parasitology, Institute of Medical Sciences, Shanghai Jiao Tong University, School of Medicine, Shanghai 200025, China*
[‡]*Chinese Center for Disease Control and Prevention, Beijing 100050, China*

**Objective.** Leptospirosis is one of the most important infectious diseases in China. A surveillance of leptospirosis at 26 surveillance spots covering 6 provinces was launched in 2005.

**Methods.** Sera of residents at these spots were collected for specific antibody detection by enzyme-linked immunosorbent assay (ELISA), and serogroup typing of the involved leptospirochetes was analyzed by

[§]Contributed equally to this work.
[¶]Corresponding author. Email: microbiology@sjtu.edu.cn

the microscopic agglutination test (MAT). The case survey included epidemic investigation, serologic confirmation and culture of pathogens. Culture of pathogens in the tissue biopsied from trapped mice at these spots was also conducted. The information including topography, geomorphology, temperature, rainfall and socioeconomic factors was considered as background data.

**Results.** The relatively high seropositive rate of the residents indicated that the human exposure to leptospira infection in these areas might be very severe. Identification of the leptospira serogroup that dominated in the involved regions revealed that there might be alteration of the serogroup of leptospires. The rates of infection of the trapped mice ranged from 0.16% to 16.26%. Infection prevalence, epidemical characteristics and the serogroup of the pathogen were also analyzed.

**Conclusion.** These results may be valuable in developing effective measures for optimizing the surveillance of leptospirosis in the epidemic areas, and in better preventing and controlling leptospirosis in both human and rodent hosts.

*Keywords*: surveillance; leptospirosis; murine reservoirs; prevention and control.

## 1. Introduction

Leptospirosis is one of the most widespread zoonoses in the world, with flu-like episodes and severe renal and hepatic damage such as hemorrhage and jaundice. It is mostly a waterborne disease, especially in rural areas,[1,2] and contact with water contaminated by urine of infected mice may be one of the major routes of infection. As agricultural regions account for the bulk of the total land area, irrigation and cultivation are very commonly seen all over China. These facilitated the prevalence of leptospirosis. In recent years, due to heavy rains and floods, especially in the provinces located along the south of Yangtze River,[3] coupled with increased water-conserving farmland construction, leptospirosis outbreaks and severe epidemics have been frequently reported, as in some other countries.[4,5] Therefore, timely collection of data and the corresponding information on leptospirosis in the endemic areas help in designing better approaches to prevention and control of the disease.

Leptospirosis being one of the statutory B infectious diseases in China, a case report was made monthly according to the data submitted by the Chinese Center for Disease Control and Prevention (China CDC) branches in each province. To increase the efficiency of interaction and communication between the public health departments and the epidemic scenes so as to acquire systemic information from epidemic investigation such as transmission patterns and the infection rate of animal reservoirs, 26 surveillance spots in high-endemic areas, including Sichuan, Jiangxi, Hunan, Hubei, Anhui and Yunnan provinces, have been selected and established since 2005. Sera of residents at these spots were collected for specific antibody detection and serogroup typing of the involved leptospirochetes.[6] A case report was made after a case survey including epidemic investigation, serologic confirmation and culture of pathogens. As rodent animals are important hosts and reservoirs in the epidemiology of leptospirosis,[7] murine hosts as major reservoirs were under surveillance in this study. The geographic information, including topography, geomorphology, temperature, rainfall and soil, along with the socioeconomic data, such as population and the demographic composition of the residents, were also included as background information. The surveillance of leptospirosis and murine reservoirs proved to be feasible and necessary.

## 2. Results

Sera of residents randomly sampled at the surveillance spots in the six provinces revealed an average positive rate of 32.22% (1129/3504) (Table 1), and the relatively high seropositive rate indicated that the human exposure to leptospira infection in these areas might be very severe.[8]

The dominant leptospira serogroups from cultured samples of the residents infected at some spots are listed in Table 2. Leptospira serogroup

**Table 1.** Seropositive rates of residents sampled from the surveillance spots, 2005.

| Province | Number of spots | Number of residents | Number of positive | Seropositive rate (%) |
|---|---|---|---|---|
| Sichuan | 5* | 505 | 346 | 68.51 |
| Jiangxi | 4 | 413 | 130 | 31.48 |
| Hunan | 4 | 423 | 294 | 69.50 |
| Hubei | 5 | 892 | 214 | 23.99 |
| Anhui | 4 | 1064 | 82 | 7.71 |
| Yunnan | 3 | 207 | 63 | 30.43 |
| Total | 25 | 3504 | 1129 | 32.22 |

*Data not obtained from one spot.

**Table 2.** Dominant serogroups of leptospires identified from cultured samples of residents at some surveillance spots, 2005.

| Province | Serogroup of leptospires (GMT or positive percentage of specific antibody by MAT) | | | | Number of samples |
|---|---|---|---|---|---|
| Sichuan | Ict (22.75) | Au (12.59) | Grip (12.53) | | 1162 |
| Jiangxi | Jav (201.29) | Can (113.14) | Au (108.34) | Bat (104.18) | 413 |
| Hunan | Ict (22.37) | Au (15.1) | Po (14.63) | Aut (14.41) | 525 |
| Hubei | Ict (46.78%)* | | | | 1117 |
| Anhui | Ict (65.85%)* | | | | 82 |

GMT — geometric mean titer; Ict — Icterohaemorrhagiae; Au — Australis; Grip — Grippotyphosa; Jav — Javanica; Can — Canicola; Bat — Bataviae; Po — Pomona; Aut — Autumnalis.
*GMT not available, by percentage.

Icterohaemorrhagiae dominated in most epidemic regions (Table 2). Alteration of the epidemic serogroup of leptospires was also observed. For example, the previous epidemic serogroup Pomona in Anhui province[3] has been replaced by serogroup Icterohaemorrhagiae.

A total of 61,626 trapping cages were placed at the surveillance spots, and 2400 wild mice were trapped, involving *Apodemus agrarius*, *Anourosorex squamipes*, *Rattus flavipetus*, *Rattus norvegicus*, *Rattus nitidus*, *Mus musculus*, *Mus minutus* and *Microtus fortis*. The species of

**Table 3.** Infection rates of murine mice trapped in surveillance spots, 2005.

| Province | Tissue samples of mice cultured | Number of positive | Rate of mice infected (%) |
|---|---|---|---|
| Sichuan | 1244 | 2 | 0.16 |
| Jiangxi | 106 | 8 | 7.6 |
| Hunan | 309 | 14 | 4.53 |
| Hubei | 513 | 8 | 1.56 |
| Anhui | 492 | 80 | 16.26 |
| Yunnan | 364 | 28 | 7.69 |

mice trapped varied according to different provinces; for example, *Anourosorex squamipes* was the dominant murine species in Sichuan, while *Rattus norvegicus* was dominant in Jiangxi and Hubei, and *Apodemus agrarius* in Anhui. The rates of infection of the trapped mice ranged from 0.16% (Sichuan) to 16.26% (Anhui) (Table 3).

## 3. Discussion

According to the data collected from China CDC for all provinces, there were 1422 leptospirosis cases, including 45 death cases reported across the nation in 2005. The morbidity rate and the mortality rate were 0.1082/100 000 and 0.0034/10 000, respectively. Patients aged from 10 to 65 accounted for 97% of the total number, with the majority being peasants (1056, 74%) related to occupational exposure and students (230, 16%).

The relatively high density and rate of infection of *Apodemus agrarius* in Anhui suggested that sufficient vigilance and measures should be taken in advance. Whether there was a relationship between the dominant murine reservoir host and the alteration of the epidemic leptospira serogroup in this area was still unknown.

There are lots of factors that affect the epidemic patterns of leptospirosis, such as regional temperature, rainfall and vegetation.[9] We analyzed the data and found that most cases (88.8%, 1263/1422) were reported during July to October. This may due to the frequent seasonal rainfall and floods in those farming areas during that period.[10]

Among the 1422 cases, 94 cases (including 2 death cases) were reported from the surveillance spots. This suggested that there might be misdiagnosed leptospirosis cases or missed cases.[11,12] In fact, the abundance of agricultural regions and the extremely complicated geographic situation in China made the surveillance of leptospirosis extremely difficult. Hence the establishment of more surveillance spots covering more endemic regions should be considered. Meanwhile, optimizing the surveillance measures and strengthening the technique collaboration, such as the access to diagnostic laboratory support between epidemic regions, should be emphasized to improve the accuracy of the monitoring and evaluation in the course of surveillance. The survey of leptospirosis in the surveillance spots may be valuable not only for establishing a more functional surveillance system, but also for formulating further measures to better reduce the risk of leptospira infection in these epidemic areas.

## Acknowledgments

This work was supported and guided by the Chinese Center for Disease Control and Prevention (China CDC), Beijing, China. We are very grateful to all the survey teams and especially to the hard-working surveyors who collected the original data from the spots in the provinces under surveillance for leptospirosis.

## References

1. Levett PN, *Clin Microbiol Rev*, 2001, **14**: 296–326.
2. Dziuban EJ *et al.*, *MMWR Surveill Summ*, 2006, **55**: 1–30.
3. Shi MH *et al.*, *Study on Geographical Epidemiology of* Leptospirosis *in China*. (Asia Medicine Press, Hong Kong, 2000), pp. 2–10.
4. Jansen A *et al.*, *Emerg Infect Dis*, 2005, **11**: 1048–1054.
5. Meites E *et al.*, *Emerg Infect Dis*, 2004, **10**: 406–412.
6. Cumberland PC *et al.*, *Am J Trop Med Hyg*, 1999, **61**: 731–734.

7. Faine S *et al.*, *Leptospira and Leptospirosis*, 2nd ed. (MediSci, Melbourne, 1999).
8. Yang WY *et al.*, *Chin J Zoon*, 2001, **17**: 111.
9. Shang Q *et al.*, *Chin J Zoon*, 1999, **15**: 95.
10. Pappachan MJ *et al.*, *J Epidemiol Comm Health*, 2004, **58**: 1054–1055.
11. Zhang JR *et al.*, *Chin J Epidemiol*, 1999, **20**: 211–214.
12. Ye CF *et al.*, *Chin J Zoon*, 2002, **18**: 110.

# 1.3

# Cloning and Expression of the Histidinol Dehydrogenase Gene from *Mycobacterium tuberculosis* H37Rv and Properties of the Recombinant Histidinol Dehydrogenase

Rui-Liang Jin*, Rui-Qing Liu*,†, Hong-Jun Wang*,†, Jian Cao†, Sheng-Feng Xu*, Li-Juan Chunyu*, Yun-Min Xu* and Hong-Hai Wang*,‡

**State Key Laboratory of Genetic Engineering, Institute of Genetics, School of Life Sciences, Fudan University, Shanghai 200433, China*
†*College of Bioengineering, Henan University of Technology, Zhengzhou 450052, China*

**Objective.** To obtain the recombinant active histidinol dehydrogenase of *Mycobacterium tuberculosis* (*Mtb*) and characterize the enzymatic properties.

**Methods.** The gene *hisD* was amplified by PCR from the genomic DNA of *Mtb* strain H37Rv and cloned into pET28a(+) to construct the recombinant expression plasmid pET-28a-HDH. The recombinant plasmid

‡Corresponding author. Email: hhwang@fudan.edu.cn

was introduced into the cells of *E. coli* strain BL21 (DE3). Recombinant MtHDH was expressed after IPTG induction and enzymatic and biophysical properties was assayed after purified with Ni-NTA resin.

**Results.** The recombinant MtHDH was purified in active state with a specific activity of 1.788 U/mg. The optimal pH and temperature of the MtHDH were 8.3 and 45°C, respectively. The relative activity was enhanced in the presence of $Mn^{2+}$, $Ca^{2+}$, $Zn^{2+}$ and $Co^{2+}$. The kinetic constants were determined: Km for $NAD^+$ was 0.9765 mmol/L and for histidinol 2.755 μmol/L. Circular dichroism studies indicated that the secondary structure of the recombinant protein had about 20.5% α-helix, 40.9% β-sheet, 4.2% β-turn and 34.3% random coil at 25°C.

**Conclusion.** The *hisD* gene of *Mycobacterium tuberculosis* H37Rv was successfully cloned and expressed. The enzymatic properties demonstrated the purified recombinant protein had activities of histidinol dehydrogenase. This work can facilitate the discovery of novel antimicrobial agents against tuberculosis and searching for immunologic epitopes.

*Keywords*: *Mycobacterium tuberculosis*; histidinol dehydrogenase; NADH; *L*-histidinol; cloning; recombinant plasmid.

## 1. Introduction

Tuberculosis (TB), caused by *Mycobacterium tuberculosis* (*Mtb*), is one of the most important human bacterial pathogens. New drugs are urgently needed to overcome some of the challenges in global tuberculosis control, such as the appearance and spread of multidrug-resistant strains and extensively drug-resistant strains, as well as tuberculosis/HIV coinfection.

*L*-histidinol dehydrogenase (HDH) is a key enzyme in histidine biosynthesis in bacteria, archebacteria, fungi and plants which is responsible for conversion from *L*-histidinol to *L*-histidine through two steps of oxidation[1–7]:

$$L\text{-histidinol} + 2NAD^+ + H_2O \rightarrow L\text{-histidine} + 2NADH + 2H^+$$

Studies on the genes of the HDH enzymes from *Salmonella typhimurium*, *Escherichia coli*, *Neurosporacrassa* and cabbage indicated that they remained highly conserved from bacteria to fungi and plants, which may serve as a drug target for designing anti-*Mtb* drugs. To initiate this study, soluble recombinant HDH (MtHDH) from *Mtb* was expressed in *E. coli*, purified, and characterized.

## 2. Results

### 2.1. *Sequence analysis of the PCR product*

The gene fragment of hisD was obtained by PCR using *Mtb* H37Rv genomic DNA as template. It was cloned into a pET-28a (+) plasmid, and then transformed into *E. coli* BL21 (DE3). The correct clone was identified by enzyme digestion and sequencing. The sequencing result was identical to the sequence of *Mtb* in the database.

### 2.2. *Expression and purification of MtHDH*

The soluble recombinant MtHDH was purified using His·Bind according to the manufacturer's protocol. The target protein was around 50 kDa in SDS polyacrylamide gel (Fig. 1). The recombinant protein eluted with 300 mmol/L imidazole was dialyzed for further study.

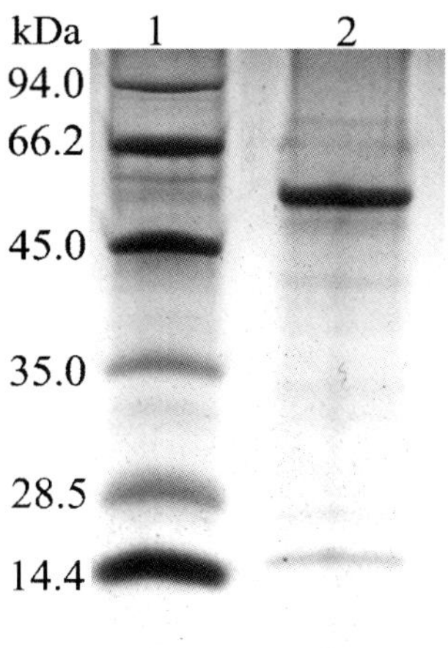


**Fig. 1.** SDS-PAGE analysis of the purified recombinant MtHDH. Lane 1 — protein molecular weight standard; Lane 2 — purified MtHDH.

## 2.3. *Enzymatic properties of MtHDH*

### 2.3.1. *Enzyme activity and specific activity of MtHDH*

The concentration of the protein was determined to be 87.34 μg/mL using the Bradford method. The specific activity of MtHDH was 1.788 U/mg according to the above-mentioned method, with three repeats. The amount of the enzyme was perfectly proportional to the reaction velocity at lower concentrations.

### 2.3.2. *Isoelectric point and optimal pH for enzyme activity*

The initial velocity was measured at different pH in a 1 mol/L NaAc/HAc buffer system (pH 4.0–6.5), a 50 mmol/L Tris-HCl buffer system (pH 6.5–9.0) and a 50 mmol/L glycine/NaOH buffer system (pH 9.0–12.0). The result showed that the optimal pH was around 8.3. The enzyme activity was lower between pH 4.0 and 8.0, and the maximal activity was achieved at pH 8.3, then it decreased fast with the further increase of pH. The relative enzyme activity at pH 11.0 was just 10% of that at optimal pH (Fig. 2).

### 2.3.3. *Effect of temperature on the enzyme activity and heat stability of MtHDH*

By testing the relative activity of MtHDH at different temperatures, it was found that the enzyme activity was the highest at 45°C (Fig. 3). At the

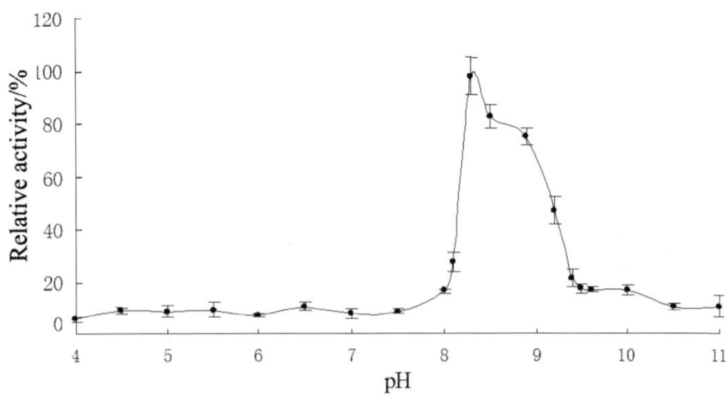


**Fig. 2.** Effect of pH on the activity of the recombinant MtHDH.

same time, it was found that the temperature had no significant effect on the enzyme activity and that the relative enzyme activity of MtHDH remained about 70% at 90°C. Treating the recombinant protein for 0–120 min at different temperatures of 70°C, 85°C and 100°C showed that the heat stability of the enzyme was quite high. After incubating for 60 min at 70°C, 85°C and 100°C, the remaining relative enzyme activities were 78.24%, 76.26% and 58.53%, respectively. Even with incubation at 100°C for 120 min, the remaining activity was 54.04% (Fig. 4).

### 2.3.4. *Effect of different ions on the enzyme activity of the recombinant MtHDH*

The relative activity of the enzyme was determined in the presence of different bivalent metal ions at 10 mmol/L in the test system. The results

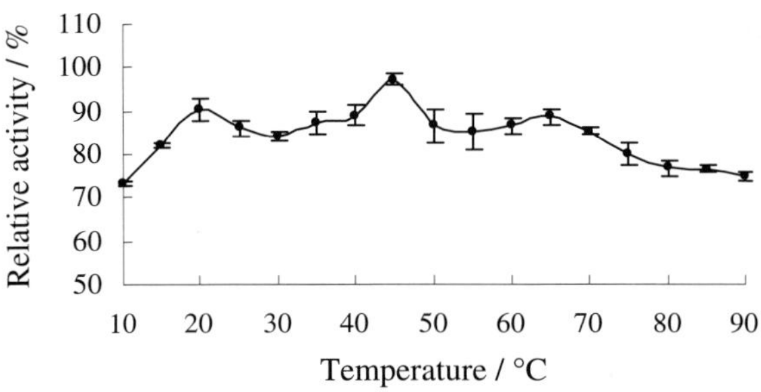


**Fig. 3.** Effect of temperature on the enzyme activity of the recombinant MtHDH.

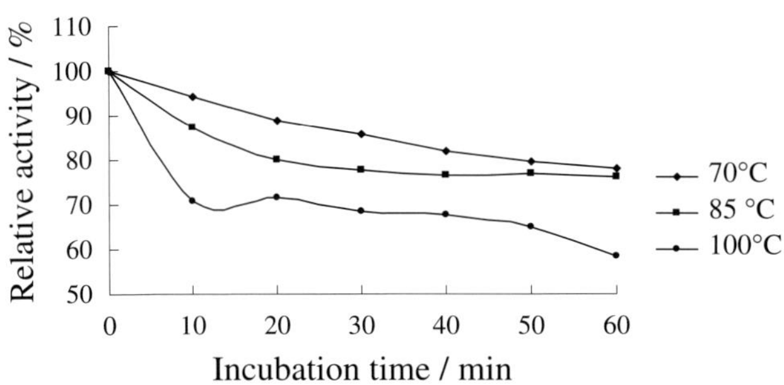


**Fig. 4.** Heat stability of the recombinant MtHDH.

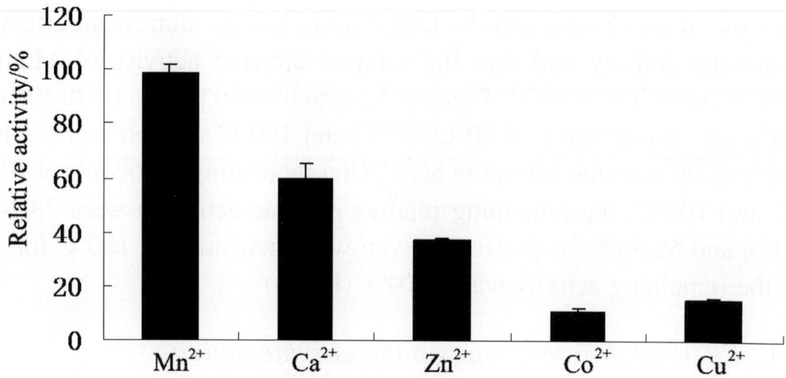


**Fig. 5.** Effect of metal ions on the MtHDH activity.

indicated that $Mn^{2+}$, $Ca^{2+}$, $Zn^{2+}$, $Co^{2+}$ and $Cu^{2+}$ activated the enzymatic reaction, while $Mg^{2+}$, $Ni^{2+}$ and $Ba^{2+}$ did not have such ability (Fig. 5).

### 2.3.5. *The Michaelis constant of MtHDH*

At the optimal reactive condition, the initial velocity ($v$) of reaction was determined using different excessive concentrations of substrate, and its reciprocal was plotted against the substrate concentration(s) (Lineweaver–Burk) to calculate the Km and $V_{max}$ value of the particular substrate. The Michaelis constants were 0.9765 mmol/L and 2.755 μmol/L for substrate $NAD^+$ and *L*-histidinol, respectively (Figs. 6 and 7).

## 2.4. *HPLC and MS analysis*

The result of HPLC assay indicated that the purity of the protein purified reached 98%. MALDI-TOF results showed that its molecular weight was 49044.0 Da (Fig. 8), with a 121.6 Da difference from its theoretic molecular weight. We thought that the recombinant protein might have lost a methionine residue (131.04 Da) during the expression process. Therefore, the difference between the theoretic molecular weight and the measured molecular weight was 9.44 Da. Data confirmed that the protein we expressed and purified was MtHDH, which was encoded by the hisD gene.

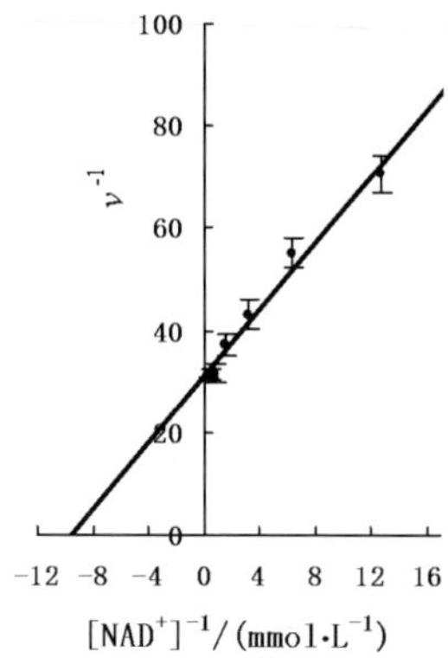


**Fig. 6.** Km and $V_{max}$ of $NAD^+$.

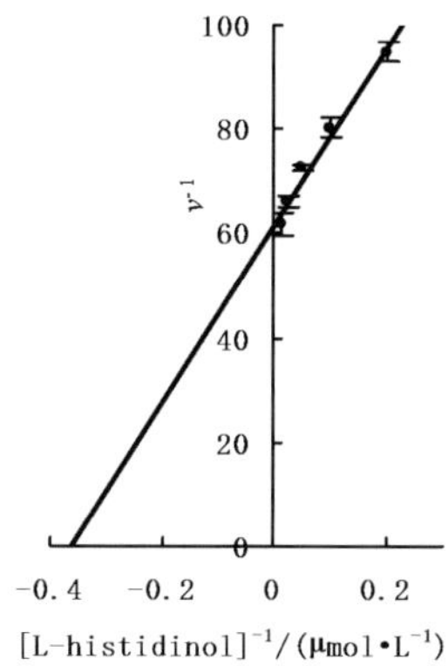


**Fig. 7.** Km and $V_{max}$ of *L*-histidinol.

## 2.5. *Preliminary assay of the MtHDH secondary structure*

By circular dichroism analysis, the MtHDH secondary structure showed the biggest ellipse degree at 195 nm, and two smallest ellipse degrees at 208 nm and 220 nm. When the result was analyzed with the PROSEC program, it was found that there was 20.5% α-helix, 40.9% β-sheet, 4.2% β-turning and 34.3% irregular curl in the MtHDH secondary structure at 25°C, which was consistent with the prediction by the PSIPRED program (http://bioinf.cs.uk/psipred). Meanwhile, the effect of temperature on the secondary structure of the protein was determined at temperatures from

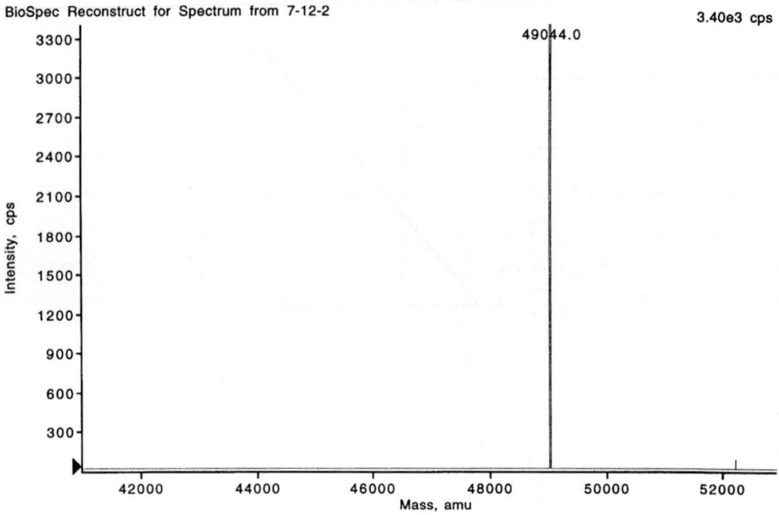


**Fig. 8.** MALDI-TOF of the mass spectrum of the purified recombinant MtHDH.

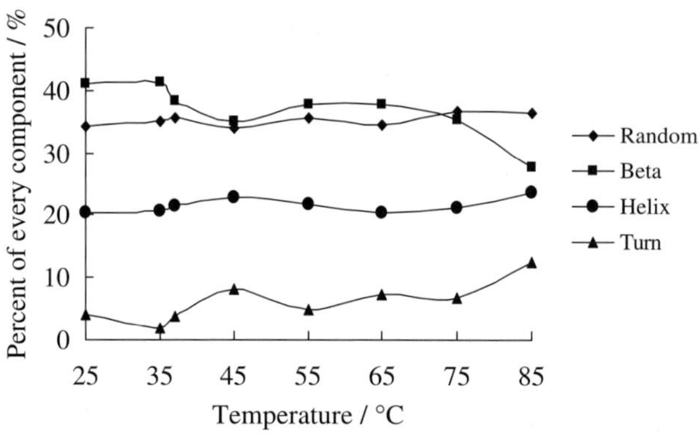


**Fig. 9.** Effect of temperature on the secondary structure of the recombinant MtHDH.

25°C to 85°C with a 10°C interval. The results confirmed that with the increase in temperature, the proportion of the α-helix and β-turning increased, while the proportion of the β-sheet decreased step by step, and the irregular curl did not change significantly (Fig. 9).

## 3. Discussion

HDH is a key enzyme in the biosynthesis of *L*-histidine in *Mtb*. Since *Mtb* grows very slowly, it is very difficult to obtain natural HDH from the bacteria. Here, we cloned the gene encoding HDH and obtained the soluble and biologically active recombinant enzyme, using genetic engineering, and further characterized its enzymological properties. During this study, we have attempted to construct two recombinant plasmids using expression vectors pET-28a and pET-30a, discovering that soluble rMtHDH was produced more efficiently using the pET-28a plasmid. Besides, we found that more soluble MtHDH was expressed at lower IPTG concentration and low temperature.

When studying the enzyme from *Salmonella typhimurium*, Charles *et al.*[8] reported that HDH is a homodimer which contains $Zn^{2+}$. HDH lost its activity when denatured by urea, or by removing $Zn^{2+}$ with a chelating reagent. The catalytic activity was restored when $Zn^{2+}$ was added. No activity was detected without a metal ion. It is possible that MtHDH lost the chelated metal ion in the process of purification, resulting in the deprivation of enzyme activity. In this study, we found that the enzymatic activity was higher in the presence of $Mn^{2+}$ than $Zn^{2}$, which was consistent with that of HDH in *E. coli*. It is predicted that most likely $Zn^{2+}$ will help to generate precipitation at higher pH and does not enter the central part of the enzyme.

Lindsay *et al.*[9] found that the optimal temperature of HDH from five psychrophilic, mesophilic and thermophilic bacteria was 24°C, 41°C, 74°C, 85°C and 92°C, respectively, which was correlated with the suitable temperature for growth of each species of bacteria. *Mtb* belongs to the mesophile, whose suitable temperature for growth is 45°C. This is consistent with Lindsay's research. The preliminary determination of rMtHDH's secondary structure indicated that each composition of secondary structure was relatively independent of temperature. We think that rMtHDH showed only limited heat denaturation with the increase in temperature, and this enzyme has high heat stability.

Helmut and Creaser[1,3] reported that the Michaelis constants of HDH from *S. typhimurium* and *N. crassa* using $NAD^+$ and *L*-histidinol as substrate were 0.7 mmol/L, 0.13 mmol/L and 14 μmol/L, 8.2 μmol/L, respectively. These are of the same order of magnitude as our result of 0.9765 mmol/L, 2.755 μmol/L. Besides, Bitarr's result indicated that the specific activity of the enzyme of *S. typhimurium* and *E. coli* was 12.5 and 12.71 U/mg,[10] which are also of the same order of magnitude as our result of 1.788 U/mg.

This study has set up a basis for subsequent structural characterization of this protein, aiming at future screening for drug targets against tuberculosis, and searching for immunologic epitopes.

## References

1. Helmut G, Biochem J, 1979, **81**: 153–157.
2. Andorn N *et al.*, *J Gen Microbiol*, 1982, **128**: 579–584.
3. Creaser EH *et al.*, *Biochem J*, 1967, **103**(1): 36–41.
4. Helmut G *et al.*, *Eur J Biochem*, 1985, **150**: 305–308.
5. Kheirolomoom A *et al.*, *Arch Biochem Biophys*, 1994, **312**(2): 493–500.
6. Atsuko N *et al.*, *Arch Biochem Biophys*, 1991, **1**: 127–132.
7. Adams E *et al.*, *J Biol Chem*, 1955, **217**: 325–344.
8. Charles G *et al.*, *Arch Biochem Biophys*, 1989, **272**(2): 311–317.
9. Lindsay JA *et al.*, *Biochem J*, 1977, **165**: 247–253.
10. Bitar KG *et al.*, *Biochim Biophys Acta*, 1977, **493**: 429–440.

# 1.4

# Cloning and Sequence Analysis of the *fbps* Gene of *Streptococcus suis* Type 2

Li-Yun Sun, Hong-Jie Fan and Cheng-Ping Lu*

*Key Laboratory of Animal Disease Diagnostic and Immunology, Ministry of Agriculture, Nanjing Agriculture University, Nanjing 210095, China*

**Objective.** To define whether *Streptococcus suis* type 2 (SS2) Jiangsu isolate HA9801 possesses the *fbps* gene [encoding fibronectin and fibrinogen-binding protein (FBPS)] and to study its sequence.

**Methods.** The *fbps* gene was amplified from genomic DNA of SS2 Jiangsu isolate HA9801 by polymerase chain reaction (PCR). The PCR product was cloned into the cloning vector pMDT-18. The recombinant fragment of 1938 bp was compared to the same gene in SS2 isolates of the Netherlands strain as reported by Greeff, and to that of the ATCC strain, and similar genes from other *Streptococci*.

**Results.** FBPS displayed 76.0%, 73.0%, 72.4%, 70.1% and 68.8% identity with FbpA of *S. gordonii* (CAA4628.2), putative fibronectin/ fibronectin-binding protein of *S. mutant* (AAN59018.1), PavA of *S. pneumoniae* (AAK75087.1), adherence and virulence protein A of

*Corresponding author. Email: lucp@njau.edu.cn

*S. agalactiae* 2603V/R (AAN00072.1) and Fbp54 of *S. pyogenes* (AAA57236), respectively.

**Conclusion.** Based on sequence analysis, FBPS was predicted as an adhesion without an anchoring component.

*Keywords*: *fbps* gene; *Streptococcus suis* type 2; clone; sequence analysis.

## 1. Introduction

Fibronectin (Fn), a heterodimer glucoprotein with a size of 440 kDa, is generated by various types of cells and exists in soluble form in plasma, humor as well as extracellular matrix (ECM). Meanwhile, there are other components such as fibrinogen (Fgn) besides Fn in ECM. Both Fn and Fgn are used as substrates for adherence by bacteria.[1] Adherence is a prerequisite for bacteria colonization[2] and successful initiation of infection.[3] It has been shown that the interactions between Fn and Fn-binding protein help *Streptococcus suis* to adhere to tissues and invade cells of the host.[4]

*S. suis*, a pathogen of a variety of infectious disease syndromes, can induce arthritis, pneumonia, meningitis and septicemia in pigs, and even meningitis in human beings.[5,6] To date, 35 serotypes of *S. suis* have been identified, and the virulence varies among different serotypes and even among different strains within the same serotype. *S. suis* type 2 is considered to be the most virulent serotype, and is also most frequently isolated from diseased animals.[6]

*S. suis* naturally colonizes in the upper respiratory tract (especially in the tonsil and nasal cavity), genital tract and alimentary canal. Charland and colleagues[7] have verified that it can adhere to epithelial cells of blood capillaries in the cerebrum. Zeng and Lu[8] have confirmed that *S. suis* type 2 isolated from Jiangsu province can adhere to epithelial cells of the tonsils, vascular endothelial cells of the lung and spleen of young rabbits as well as HEp-2 *in vitro*, and that the muramidase-released protein (MRP) plays a role in bacteria adherence. Nevertheless, whether there are other adhesion molecules besides MRP needs to be further studied. Greeff and colleagues[9] lately reported that the Fn- and Fgn-binding proteins

(FBPS) in *S. suis* type 2 have virulence effects. No related research in China is available at present. It deserves to be studied whether there is an *fbps* gene in the Chinese isolate of *S. suis* type 2 or not, and whether there is any difference between the genes from domestic and foreign isolates. Therefore, we cloned and sequenced the *fbps* gene of *S. suis* type 2 HA9801 isolated form Jiangsu province. The work will facilitate future research on the function and application of FBPS.

## 2. Results

### 2.1. *PCR amplification*

The primers designed based on the *fbps* gene sequence of *S. suis* type 2 (SS2) strain 10 recorded in GenBank successfully amplify a specific band from the HA9801 strain of SS2. The PCR product showed a band about 1.9 kb in size in agarose electrophoresis, which is identical to the expectation.

### 2.2. *Construction of a recombinant plasmid*

The PCR product was purified and legated to the pMD-T vector to construct a recombinant plasmid. Then the ligated product was transformed into DH5α. The recombinant plasmid was extracted and digested with restriction enzyme *Bam* HI. A single band larger than the vector digested by the same enzyme was seen in electrophoresis.

The sequence, which contains a 1938 bp ORF, was confirmed as an *fbps* gene after sequencing and analyzing with the BLAST software. The *fbps* gene sequence reported in this study was submitted to GenBank with accession number AY566303.

### 2.3. *Analysis of the* **fbps** *gene sequence*

The homology between the *fbps* gene cloned in this study and that of the SS2 strain isolated from the Netherlands was found to be more than 99.99% using the DNAstar program. There are five differences in sites 82, 292, 905, 939 and 967, as A, A, G, G, C in the HA9801 strain and C, C, A, A, G in the Netherlands strain. There are three differences between the *fbps* sequence cloned in this study and ATCC43765 in sites 292, 905 and 1512, as A, G, A in HA9801 and G, A, G in ATCC.

## 2.4. *Deduced FBPS protein sequence and analysis of the protein structure*

The cloned *fbps* gene was predicted to encode a protein with 552 amino acids by DNAstar, of which there are 57 leucines, 47 glutamic acids and 47 lysines. Four differences were found between the cloned *fbps* sequence and the SS2 strain isolated from the Netherlands in sites 28, 98, 302 and 323, as N, S, G, L in the HA9801 strain and H, R, E, V in the Netherlands strain (Fig. 1). There were two differences between the protein sequence of ATCC43765 and that of the HA9801 strain in sites 98 and 302, as S, G in the HA9801 strain and R, E in the ATCC43765.

Analyzed by DNAstar, the FBPS protein of HA9801 was found to have a molecular weight of 63 kDa and an isoelectric point of 7.41. Of the total of 552 amino acids, 33.7% are the amino acids with electric charge, and the hydrophilic amino acids dominate. Two alpha-superhelixes were predicted between the amino acid residues from 266 to 306, and 354 to 392 of the FBPS protein of the HA9801 strain. A leucine zipper motif was found among the amino acid residues from 281 to 318. Analysis by the ProDom software (http://protein.toulouse.inra.fr/prodom/doc/prodom.html) indicated that this FBPS has two putative domains located at amino acid residues from 1 to 431 and 434 to 521, respectively. The first one is a domain similar to that of the N terminal of FbpA protein,

MSFDGFFLHHMTAELRANLEGGKIQKINQPFEQEIVLNIRSNRQSHKLLLSAHSVFGRVQLTQSDFTNPIKVPNTFTMI
LRKYLQGAIIEEIRQLLNDSILEFSVSNKDEIGDHIQATLIVEIMGKHSNIILVDKSEQKIIEATKHVGFSQNSYRTIL
PGSTYIRPPETHSLNPYTVSDEKLFEILSTQELSPKNLQQVFQGLGRDTASELANHLQIDRLKNFRAFFDQATQPSLTD
KSYAALPFANSPENQPHFESLSSLLDFYYQDKAERDRVAQQANE L IKRVASE L EKNRKK L IKQEQG L ADTETAE L VRQK
GEL L TTY LHQVPNDQSSVRLDNYYTGKELEIELDVALTPSQNAQRYFKKYQKLKEAVKHLTNLIEETKSTIVYLESVDT
MLGQASLAEIDEIREELIETGYLKRRHREKIHKRQKPERYLATDGKTIILVGKNNLQNDELTFKMAKKG ELWFHAKDIP
GSHVVITDNLDPSDEVKTDAAELAAYFSKARHSNLVQVDMIEAKKLHKPTGGKPGFVTYRGQKTLRVTPTEDKIKSMKI

**Fig. 1.** The deduced amino acid sequence of the *fbps* gene of the SS2 Jiangsu isolate. The underlined residues are different from that of SS2 strain 10. The leucine zipper is underlined with a double dash and the D/E-X-W/Y-X-H motif is underlined with a curve.

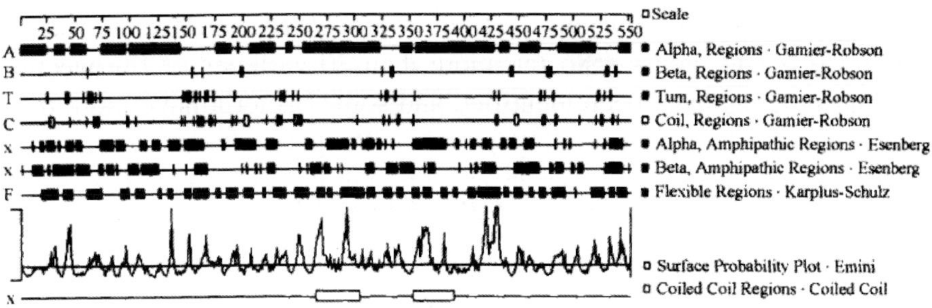


**Fig. 2.** Prediction of the structure of FBPS of the SS2 Jiangsu isolate.

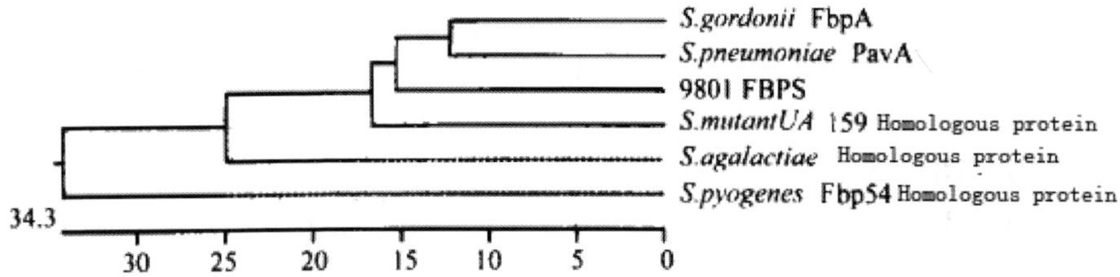


**Fig. 3.** Phylogenetic analysis of protein sequences showing the relationship of FBPS of the SS2 Jiangsu isolate to the homologous sequences from the genus *Streptococcus*.

which binds to Fn, and the second one exists in the protein notated as Fn/Fgn binding protein, which has a conserved motif of D/E-X-W/Y-X-H with a significant function (Fig. 2).

Cluster comparison using DNAstar indicates that the homology between the FBPS protein of HA9801 and FbpA of *S. gordonii*, the predicted Fn/Fgn binding protein of *Streptococcus* mutants, PavA of *S. pneumoniae*, the adhesion and virulence protein A of *S. Agalactiae* and Fbp54 of pyogenic *Streptococcus* is 76.0%, 73.0%, 72.4%, 70.1% and 68.8%, respectively. Figure 3 shows the cladogram result.

## 3. Discussion

Fibronectin-binding proteins are important bacterial afimbrial adhesins. Till 2002, 16 species of Gram-positive bacteria that express fibronectin had been identified, including pyogenic *Streptococcus*, *Staphylococcus aureus*, *Borrelia burgdorferi*, *Listeria monocytogenes*, *Mycobacterium tuberculosis* and others.[10] So far, more than 20 varieties of fibronectin-binding proteins have been identified and some bacteria have more than one kind of fibronectin-binding proteins. Pyogenic *Streptococcus* was found to have 12 different fibronectin-binding proteins, including Sfb I, F2 protein, Sfb II, M1, Sof22, GPDH, Fbp54 and so on.[11]

Till now, fibronectin-binding proteins of SS2 have been reported only by Greeff *et al.*[9] and therefore deserve further study. In this study, we cloned the *fbps* gene from SS2 isolated from Jiangsu province and found that it is highly homologous between domestic and foreign isolated strains.

The ORF included in the *fbps* gene of SS2 encodes a protein with 552 amino acids, which is the same as the fibronectin-binding proteins Fbp54 of pyogenic *Streptococcus*, PavA of *S. pneumoniae*, FbpA of *S. gordonii* and the adhesion and virulence protein A of *S. agalactiae*, in that they have no N-terminal signal peptides sequence[12] and no C-terminal LPXTG sequence for cell wall anchoring of gram-positive bacteria.[13] It has been confirmed that PavA is an anchorless adhesin[14] and the structure of FBPS is similar to that of PavA. Therefore we conjecture that FBPS is also an anchorless adhesin.

The molecular weight of FBPS of the Netherlands isolate, as reported by Greeff, is 63.3 kDa, and the isoelectric point is 7.43. Based on the *fbps* sequence cloned by the author, the molecular weight and isoelectric point of the ATTC43765 strain are postulated to be $63.3 \times 10^3$ Da and 7.41, respectively. The FBPS proteins of the Netherlands strains derived from pigs and the ATCC43765 strain are basically identical with the strain isolated from pigs in Jiangsu province in both the secondary and hydrophobic structure, and the distribution of epitopes. They were all predicted to have two alpha-superhelixes between the amino acid residues from 266 to 306, and from 354 to 392. This similarity indicates that the FBPS from different areas and different isolates have no significant difference between each other.

FBPS have repeat sequence,[11] which is a characteristic shared by all the other surface proteins of gram-positive bacteria. It has been considered by some people that the repeat sequence plays an important role in the immune escape through antigen variation.[14] Besides, Greeff *et al.*[9]

have confirmed that the *fbps* gene exists in all kinds of serotype SS2 except 32 and 34. The convalescent serum of infected animals has ingredients that have reacted with FBPS. The mice systematically and mucosally immunized by Fbp45 induced protective immune response.[15] And the structure of FBPS is highly homologous to that of the adhesin Fbp45 of pyogenic *Streptococcus*. It is not difficult to imagine that *fbps* and FBPS can be applied in the detection and immunization of *S. suis*.

## References

1. Roumen P *et al.*, *J Cell Sci*, 2002, **115**: 3861–3863.
2. Lu CP, *Veterinary Microbiology*, 3rd ed. (China Agricultural Press, Beijing, 2001), pp. 58, 336.
3. Beachey EH *et al.*, *J Infect Dis*, 1981, **143**: 325–344.
4. Danny J *et al.*, *Matrix Biol*, 1999, **18**: 211–223.
5. Arends JP *et al.*, *Rev Infect Dis*, 1988, **10**: 131–137.
6. Allgaier A *et al.*, *G J Clin Microbiol*, 2001, **39**: 445–453.
7. Charland N *et al.*, *Infect Immun*, 2000, **68**: 637–643.
8. Zeng QY *et al.*, *J Nanjing Agr Univ*, 1999, **22**: 67–71.
9. Greeff A *et al.*, *Infect Immun*, 2002, **70**: 1319–1325.
10. Navarre WW *et al.*, *Microbiol Mol Biol Rev*, 1999, **63**: 174–229.
11. Yother J *et al.*, *J Bacteriol*, 1994, **176**: 2976–2985.
12. Whatmore AM *et al.*, *Microbiology*, 2001, **147**: 419–429.
13. Holmes AR *et al.*, *Mol Microbiol*, 2001, **41**: 1395–1408.
14. Shigetada K *et al.*, *Infect Immun*, 2001, **69**: 924–930.
15. Kreikemeyer B *et al.*, *Int J Med Microbiol*, 2004, **294**: 177–188.

**1.5**

# Effects of EnvZ Histidine Kinase Inhibitors on the Virulence of *Shigella flexneri*

Ming-Liang Chen*, Qi Wang[†], Sheng-Fu Dong*, Zhi-Qiang Qin*, Jian Zhang[†], Cong Han*, Xia Cai[†], Xu Shen [†] and Di Qu*,[‡]

**Key Laboratory of Medical Molecular Virology, Shanghai Medical College, Institutes of Biomedical Sciences, Fudan University, Shanghai 200032, China*

[†]*Drug Discovery and Design Center, State Key Laboratory of Drug Research, Shanghai Institute of Materia Medica, Chinese Academy of Sciences, Shanghai 201203, China*

**Objective.** To explore the effects of EnvZ histidine kinase inhibitors on the virulence of *Shigella flexneri*.

**Methods.** The recombinant *S. flexneri* SF-EnvZc was expressed and purified in the *Escherichia coli* system. By using the Kinase-Glo$^{TM}$ luminescent kinase assay to screen against the small chemical library, compounds that inhibit the autophosphorylation activity of SF-EnvZc were selected as EnvZ inhibitors. The binding affinities of the inhibitors to SF-EnvZc protein were determined by using surface plasmon resonance (SPR) *in vitro*. The mouse Sereny test was used

[‡]Corresponding author. Email: dqu@shmu.edu.cn

to evaluate the effects of the EnvZ inhibitors on the virulence of *S. flexneri*.

**Results.** Compounds 1, 2 and 3 could inhibit the activity of SF-EnvZc autophosphorylation with $IC_{50}$ values of 70.17 ± 5.83 μmol/L, 10.18 ± 0.86 μmol/L and 37.66 ± 9.64 μmol/L, respectively. Their binding affinities to SF-EnvZc were evaluated with $K_D$ values of 4.08 μmol/L, 3.57 μmol/L and 2.94 μmol/L, respectively. In the mouse Sereny test, mice inoculated with *S. flexneri* strain 9380 and treated with compounds 1 and 2 displayed no disease while those treated with compound 3 displayed a slight conjunctival inflammation (+) in comparison with a DMSO-treated control group, in which a severe keratoconjunctival inflammation (+++) was observed. The minimal effective concentrations of compounds 1, 2 and 3 to release mouse Sereny reaction were 12.5 μmol/L, 12.5 μmol/L and 100 μmol/L, respectively.

**Conclusion.** The three selected EnvZ inhibitors relieved mouse Sereny reaction of *S. flexneri*, suggesting that inhibition of EnvZ could reduce the virulence of *S. flexneri*.

*Keywords*: *Shigella flexneri*; histidine kinase; EnvZ; surface plasmon resonance.

## 1. Introduction

Bacterial two-component signal transduction systems (TCSTSs), generally composed of a histidine protein kinase and a response regulator protein, were found in most prokaryotes but not in mammals. A histidine protein kinase, which responds to changes of environmental conditions, undergoes ATP-dependent autophosphorylation on a specific histidine residue and subsequently transfers the phosphoryl group to an aspartate residue on a cognate response regulator protein. This results in regulation of various functions, including sporulation, osmoregulation, chematropism, photosynthesis and pathogenicity.[1] Currently, bacterial TCSTSs are attracting increasing attention due to their roles as potential novel antibacterial targets, especially in those which are required for bacterial growth and virulence in pathogenic microorganisms.[2] Recently Watanabe *et al.*[3] chose YycG/YycF — a histidine protein kinase of a TCSTS which can regulate bacterial growth in *Bacillus subtilis* and *Staphylococcus aureus* as a target of antimicrobial drugs and screened for new inhibitors. Therefore, we investigated EnvZ/OmpR — a TCSTS of *Shigella* — as a drug target. In the EnvZ/OmpR system, EnvZ senses the extracellular osmotic pressure and regulates the level of phosphorylated OmpR.[4] The OmpR plays an important part in regulating the virulence of *Shigella*.[5] The knockout mutant of *envZ-ompR* genes from *Shigella flexneri* M90T showed decreased virulence, which suggested that EnvZ is a potential antibiosis target.[5]

In this work, we chose the EnvZ HATPase C domain (substrate binding region) of *S. flexneri* as the potential target to screen for candidate inhibitors from the chemical library and detected the effect of candidate inhibitors on the virulence of *Shigella*, aimed at exploring leading compounds of EnvZ inhibitors that may be applied in treatment of *Shigella* infection.

## 2. Results

### 2.1. *Expression and purification of recombinant SF-EnvZc*

In order to investigate the effects of the compounds on enzymatic activities of SF-EnvZc, the recombinant His-tagged SF-EnvZc was overexpressed, purified, and analyzed by SDS-PAGE (Fig. 1A) and Western blot (Fig. 1B) with a mouse anti-His-tag monoclonal antibody. The molecular mass of the recombinant SF-EnvZc is 27.0 Da, with a purity > 90%.

### 2.2. *$IC_{50}$ values of candidate SF-EnvZc inhibitors*

Three compounds, called compounds 1, 2 and 3 (Fig. 2), showed inhibitory effects on enzyme activities of SF-EnvZc. The $IC_{50}$ values of compounds 1–3 were 70.17 ± 5.83 μmol/L, 10.18 ± 0.86 μmol/L and 237.66 ± 9.64 μmol/L, respectively (Table 1).

### 2.3. *Binding affinities of candidate compounds to SF-EnvZc*

In order to confirm the interaction between the potential EnvZ inhibitors and their putative target protein, the binding affinities of the three EnvZ

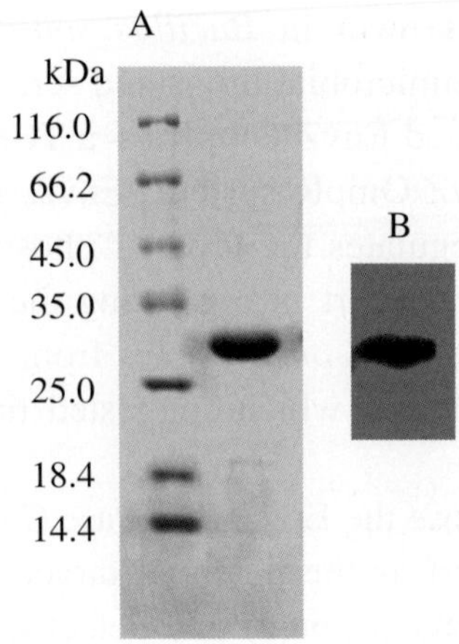


**Fig. 1.** Recombinant SF-EnvZc was identified by SDS-PAGE (A) and Western blot (B). The identity of the purified protein was confirmed by Western blot using a monoclonal antibody directed against the His tag. The molecular weight of the protein is 27.0 kDa.

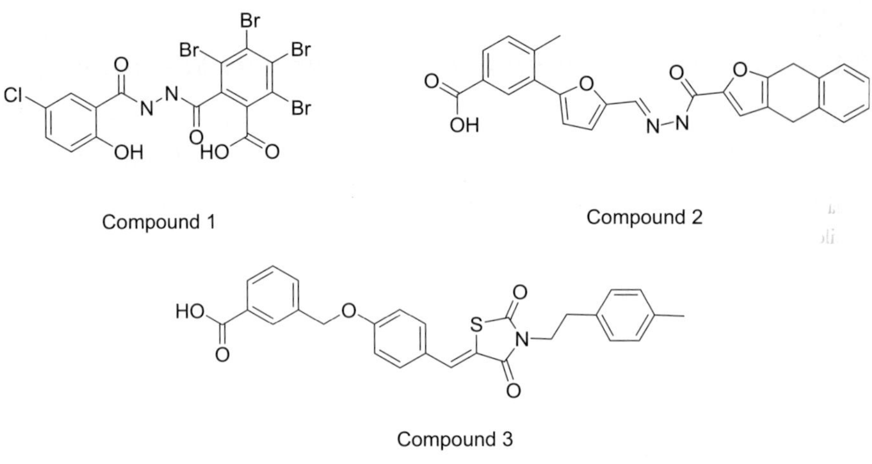


**Fig. 2.** The chemical structures of three compounds as EnvZ inhibitors.

inhibitors to the SF-EnvZc protein were determined by using SPR *in vitro*. All the three compounds (1–3) displayed high binding affinities to the SF-EnvZc protein, with $K_D$ values of 4.08 μmol/L, 3.57 μmol/L and 2.94 μmol/L, respectively (Table 1), in a dose-dependent binding pattern.

**Table 1.** Biological effects of three EnvZ histidine kinase inhibitors.

| Compound (μmol/L) | $IC_{50}$ (μmol/L) for SF-EnvZc[1] | $K_D$ value (μmol/L)[2] | $CC_5$ on Vero cell[3] |
|---|---|---|---|
| 1 | 70.17 ± 5.83 | 4.08 | >200 |
| 2 | 10.18 ± 0.86 | 3.57 | >200 |
| 3 | 37.66 ± 9.64 | 2.94 | >200 |

[1] $IC_{50}$ represents the concentration of inhibition of 50% of the SF-EnvZc protein autophosphorylation; [2] $K_D$ represents the equilibrium dissociation constant; [3] $CC_{50}$ represents the concentration that produces a 50% cytotoxicity effect on the Vero cell, for the highest concentration tested was up to

**Table 2.** Mouse Sereny test with the *Shigella* inoculated with EnvZ inhibitors.

| Inoculated compound[1] | Concentration (μmol/L) | | | | | | Control | |
|---|---|---|---|---|---|---|---|---|
| | 200 | 100 | 50 | 25 | 12.5 | 6.25 | ATCC 25922[3] | *Sf* 9380[4] |
| DMSO[5] | +++[2] | +++ | +++ | +++ | +++ | +++ | – | +++ |
| 1 | – | – | – | ++ | ++ | +++ | – | +++ |
| 2 | – | – | ++ | ++ | ++ | +++ | – | +++ |
| 3 | + | ++ | +++ | +++ | +++ | +++ | – | +++ |

[1] *Shigella flexneri* 9380 treated with each concentration of the compound was inoculated into one eye of each of six mice; [2] Sereny reaction was rated as follows: –, no disease or mild irritation; +, mild conjunctivitis or late development and/or rapid clearing of symptoms; ++, keratoconjunctivitis without purulence; +++, fully developed keratoconjunctivitis with purulence; [3] ATCC 25922, an avirulent *E. coli* strain, served as the negative control of the mouse Sereny test; [4] *Sf* 9380, a *Shigella* clinical strain with strong virulence, served as the positive control of the mouse Sereny test; [5] DMSO is the solvent of the compounds; *Sf* 9380 treated with the same amount of DMSO as in other groups served as the blank control of the mouse Sereny test.

## 2.4. *Mouse Sereny test*

To test the virulence of *S. flexneri*, the mouse Sereny test was used.[6] Compounds 1 and 2 obviously inhibited mouse keratoconjunctivitis inflammation (from +++ to –) with *Sf* 9380, and compound 3 could decrease keratoconjunctivitis inflammation (from +++ to +) (Table 2). The minimum effective drug concentrations of the three compounds were

12.5 μmol/L, 12.5 μmol/L and 100 μmol/L, respectively. Mice inoculated with *sf* 9380 displayed a severe keratoconjunctival inflammation (+++), while the control group inoculated with ATCC 25922 displayed no inflammation (–).

## 3. Discussion

*Shigella* species were known to spread directly through the fecal–oral route and invade mucous membrane of the colon. As few as 10–100 organisms can cause shigellosis,[7] which is prevalent in the developing world. It was estimated that 1.1 million deaths and 160 million cases annually were attributed to shigellosis, most common in children less than 5 years old in developing countries.[8] *S. dysenteriae* caused outbreaks of dysentery in mid-Africa, Southeast Asia and India, while *S. sonnei* could cause epidemics of dysentery in developed countries. Although the clinical manifestations of bacillary dysentery ranged from mild watery diarrhea to severe dysentery with frequent passage of bloody, mucoid, small-volume stools in humans, *Shigella* infections could cause destruction of colon epithelium and result in acute colitis.

There have been remarkable increases of antibiotic resistance in *Shigella* isolates, although some of the *Shigella* strains are still sensitive to sulfonamides, phytomycin and chloramphenicol. *Shigella* isolates resistant to antimicrobial drug sulfonamides, decycline, chloramphenicol and ampicillin are increasing at a fast speed and there is a broad spectrum of drug resistance.[9] New therapeutic targets and drugs are thus urgently needed to decrease the incidence of shigellosis worldwide.

Bernardini *et al.*[5] constructed the *envZ-ompR* mutants in *Shigella* virulent strain M90T and, with the mouse Sereny test, HeLa cell invasion test and plaque assay, they proved that the virulence of this mutant strain decreased markedly. With reverting mutation, the virulence reversed, which proved that the TCSTSs' EnvZ/OmpR could positively regulate the virulence of *Shigella*. Based on those reports, we selected EnvZ as a target for screening compounds against *Shigella*. Analysis of amino acid sequences showed that EnvZ protein was extremely conserved, and there was 100% of homology, among several *Shigella* strains (*S. dysenteriae* 197, *S. flexneri* 301, *S. boydii* 227, *S. sonnei* 046). Therefore we selected

EnvZ protein of the *Sf* 301 strain to construct prokaryotic expression plasmid, expressed and purified SF-EnvZc protein. In order to screen compounds that may inhibit SF-EnvZc phosphorylation as candidates for EnvZ inhibitors, we used Kinase-Glo™ luminescent kinase assay to detect the inhibitory effects of compounds on autophosphorylation of SF-EnvZc. The principle of the method is that the HATPase C domain of histidine protein kinase has ATP binding activity and can be autophosphorylated and the luminescent substrate can bind to ATP remaining in the solution. There was a linear relationship between the luminescent signal and the amount of ATP in the kinase reaction used in the Kinase-Glo™ luminescent kinase assay. This quantification of the relative fluorescence unit (RLU) could indirectly reflect the biological activity of SF-EnvZc. By comparing the RLU before and after these compounds were added, the inhibition of ATP binding to SF-EnvZc, indicated by the inhibition of SF-EnvZc autophosphorylation, could thus be assayed. In this study, the binding affinities of compounds 1–3 to SF-EnvZc *in vitro* were determined by employing SPR. Since HATPase C domains are substrate binding regions where EnvZ protein binds with ATP and undergoes autophosphorylation, EnvZ inhibitors can bind to the substrate region and block ATP binding. SPR results showed that compounds 1–3 could bind with SF-EnvZc in high affinities. Based on these results, we predicted that these compounds might inhibit the function of SF-EnvZc, though no effect on *Shigella* growth was observed by these compounds. All three EnvZ inhibitors were effective by decreasing the mouse Sereny reaction of *sf* 9380 to different levels (Table 1), indicating that they could inhibit the virulence of *Shigella*. This is in accordance with findings of Bernardini *et al.*, showing that EnvZ regulated the virulence of *Shigella*.

In TCSTSs' EnvZ/OmpR, EnvZ regulates OmpR by sensing the extracellular osmotic pressure,[4] and OmpR regulates several virulence proteins (such as VirR) and outer membrane pore protein OmpF–OmpC, among which VirR participates in invading endothelial cells by the bacteria.[5] OmpF–OmpC was reported to be related to *Shigella* amplification in cells.[10] The possible mechanism for inhibitors of EnvZ affecting the virulence of *Shigella* could be that by binding to *Shigella* EnvZ protein the EnvZ inhibitors inhibited the OmpR phosphorylation and downregulated the expression of virulence proteins VirR and OmpF–OmpC, thus decreasing the virulence of *Shigella*.

Studies are ongoing to further clarify the possible biochemical and molecular mechanisms of this inhibition by the virulence of *Shigella.*

## References

1. Appleby JL *et al.*, *Cell*, 1996, **86**: 845–848.
2. Stephenson K *et al.*, *Curr Drug Targets Infect Disord*, 2002, **2**: 235–246.
3. Watanabe T *et al.*, *J Antibiot* (*Tokyo*), 2003, **56**: 1045–1052.
4. Yoshida T *et al.*, *J Biol Chem*, 2006, **281**:17114–17121.
5. Bernardini ML *et al.*, *J Bacteriol*, 1990, **172**: 6274–6281.
6. Murayama SY *et al.*, *S Infect Immun*, 1986, **51**: 696–698.
7. DuPont HL *et al.*, *J Infect Dis*, 1989, **159**: 1126–1128.
8. Kotloff KL *et al.*, *Bull World Health Organ*, 1999, **77**: 651–666.
9. Chu YW *et al.*, *Antimicrob Agents Chemother*, 1998, **42**: 440–443.
10. Ferrario M *et al.*, *J Bacteriol*, 1995, **177**: 103–113.

# 1.6

# Detection of Virulence and Genes for Viable but Nonculturable (VBNC) *Campylobacter jejuni*

Jie-Lin Yang*,§, Tong-Hai Dou†, Ming Gu* and Zhong-Liang Wu‡

**Shanghai Entry and Export Inspection and Quarantine Bureau, Shanghai 200135, China*
†*Institute of Genetics, Fudan University, Shanghai 200433, China*
‡*Biomerieux Industry China, Shanghai 200001, China*

**Objective.** The existence of a viable but nonculturable (VBNC) state has been described for *Campylobacter jejuni*, and we developed a method for detecting potential virulence factors on VBNC state established VeroE6 and HeLa cell models.

**Methods.** VBNC status was induced by storing *C. jejuni* at 0°C, –70°C, and starvation. Cytotoxic distending toxin (CDT) virulence assay was used to detect VBNC *C. jejuni* in HeLa and Vero cell models, and the results were further confirmed by the RT-PCR test. VBNC *C. jejuni* remained virulent to cell cultures, though the virulence was decreased. Total and active cells were counted by conventional culture and 4′6-diamidino-2-phenylindole (DAPI) staining. CTC and DAPI were used as indirect markers for confirming the presence of VBNC *C. jejuni.*

§Corresponding author. Email: yangjl@shciq.gov.cn

**Results.** The presence of VBNC cells was confirmed by different cell count methods. *C. jejuni* decreased by about one order of magnitude after refrigeration. Before refrigeration the virulence of *C. jejuni* on cells was potent; around 70% of cells progressed to apoptosis at 72 h, while only 40% progressed to apoptosis after *C. jejuni* was refrigerated.

**Conclusion.** Low temperature and poor nutrients could induce *C. jejuni* into the VBNC state. Though remarkable decrease of VBNC *C. jejuni* virulence was observed, the VBNC *C. jejuni* could still be virulent to cells.

*Keywords*: *Campylobacter jejuni*; viable but nonculturable; cell count; enterotoxity.

## 1. Introduction

*Campylobacter jejuni* was successfully isolated and characterized in 1973.[1] It can cause human cramps, abdominal pain, profuse diarrhea, headache, malaise and fever. *Campylobacter* can also infect domestic animals, causing necrotic hepatitis to chickens and other diseases such as diarrhea of chickens, calves and piglets. Siddique *et al.* first reported its enterotoxin in 1983.[2] Later, it was found that *C. jejuni* can also produce cytotonic enterotoxin (CE), cytotoxin c and cytolethal distending toxin (CDT). Assays using cells have been widely employed for the evaluation of CE, cytotoxin c and CDT virulence. Appropriate cell lines were chosen to detect various toxins according to the sensitivity of different toxins in different cells.[3,4]

After studying the survival status of *Vibrio cholerae* and *Escherichia coli* in 1982, Xu *et al.* proposed a viable but nonculturable (VBNC) state of these bacteria.[5] Rollins *et al.* reported the existence of a *C. jejuni* VBNC state and the conditions for inducing *C. jejuni* into the VBNC state, as well as the toxin gene expression in the laboratory.[6] Because of the effects of low temperature, draught and other physical and chemical factors, many bacteria can exist in the VBNC state; however, the conventional culture method cannot detect VBNC *C. jejuni* or assess whether it retains toxicity.

The objective of this study is to detect the production of enterotoxin by means of cytological experiments and molecular methods of RNA assays after *C. jejuni* has been induced into the VBNC state. In addition, experiments on the effects of low temperature on enterotoxin production and its virulence were conducted to evaluate the toxicity of *C. jejuni* in frozen foods.

## 2. Results

### 2.1. *Bacterial counts*

*C. jejuni* growth in the VBNC state was observed by counting the total bacterial number after incubation with M199 and the viable bacterial count after staining with DAPI (Tables 1 and 2). Table 1 shows that after freezing at 70°C, the number of viable bacteria decreased from $10^6$/ml to $10^3$/ml, while with refrigeration at 4°C a two-orders-of-magnitude decrease of viable bacteria happened.

### 2.2. *Cells for detection of enterotoxin*

As shown in Table 3, in HeLa cells, after addition of the supernatant from bacteria cultured at 37°C, the cell shape gradually became longer,

**Table 1.** *C. jejuni* before and after refrigeration.

| | *C. jejuni* counts (42°C) | | Plate counts after refrigeration | |
|---|---|---|---|---|
| 4°C | $1.2 \times 10^6$/ml | $1.6 \times 10^6$/ml | $1.3 \times 10^4$/ml | $1.9 \times 10^4$/ml |
| –70°C | $1.9 \times 10^6$/ml | $1.4 \times 10^6$/ml | $6.1 \times 10^3$/ml | $6.3 \times 10^3$/ml |

**Table 2.** Number of *C. jejuni* in different states before and after refrigeration.

| 42°C culture (/ml) | | | | 4°C and –70°C refrigeration (/ml) | |
|---|---|---|---|---|---|
| | | | | 4°C | –70°C |
| Total number (mean) | $8.5 \times 10^8$ | $5.6 \times 10^8$ | $2.7 \times 10^7$ | $8.2 \times 10^6$ | |
| Total viable counts (mean) | $5.6 \times 10^8$ | $5.3 \times 10^8$ | $2.0 \times 10^7$ | $8.0 \times 10^6$ | $8.0 \times 10^5$ |
| Plate counts (mean) | $5.3 \times 10^8$ | | $1.7 \times 10^6$ | $2.5 \times 10^5$ | $3.1 \times 10^4$ |

**Table 3.** *C. jejuni* toxicity on HeLa and VeroE6 cells.

| | | Time point (h) | | | | | |
|---|---|---|---|---|---|---|---|
| | | 12 | 24 | 36 | 48 | 60 | 72 |
| HeLa 37°C | | + | ++ | +++ | ++++ | +++++ | |
| 4°C | S | +++ | +++ | ++++ | +++++ | +++++ | |
| | N | — | — | — | + | ++ | |
| −70°C | S | +++ | ++++ | +++++ | +++++ | +++++ | |
| | N | — | — | + | ++ | +++ | |
| M | | — | — | — | — | — | |
| S | | +++ | +++ | ++++ | ++++ | +++ | |
| N | | — | — | — | + | + | |
| VeroE6 37°C | | ++ | +++ | ++++ | +++++ | +++++ | +++++ |
| 4°C | S | +++ | ++++ | +++++ | +++++ | +++++ | +++++ |
| | N | — | — | + | ++ | ++ | +++ |
| −70°C | S | ++++ | +++++ | +++++ | +++++ | ++++ | +++++ |
| | N | — | + | ++ | +++ | +++ | ++++ |
| M | | — | — | — | — | — | — |
| S | | +++ | ++++ | +++++ | +++++ | +++++ | +++++ |
| N | | — | — | — | + | + | ++ |

*Qualitative standard*: HeLa: +, elongation, dark particle 20%; ++, elongation, dark particle 20–40%, round 20%; +++, elongation, dark particle 40–70%, round 20–40%, lysis 10–20%; ++++, elongation, dark particle 20%, round 40–60%, lysis 20–60%; +++++, round 40%, lysis > 60%; VeroE6: +, elongation and collapse 30%; ++, elongation and slight collapse 30–60%; +++, elongation and slight collapse 60–80%, lysis 20–40%; ++++, elongation and slight collapse 40–60%, lysis 40–60%; +++++, collapse 40%, lysis > 60%. M — M199 medium; N — salt water group; S — surface water.

and a large number of intracellular dark particles appeared, with partial disruption of cells, peaked at around 60 h; while, in VeroE6, the shape of the cultured cells, after addition of the supernatant, became slenderly disrupted and peaked at around 40 h. After the low-temperature freezing of the bacteria, although some cells had entered the VBNC state, and the toxin production capacity had declined significantly, they retained their toxin production capability, as shown by morphological change (Fig. 1).

As shown in Table 3, the M199 medium, as a blank control, had no impact on cell growth; cell growth had changed after incubation for 72 h

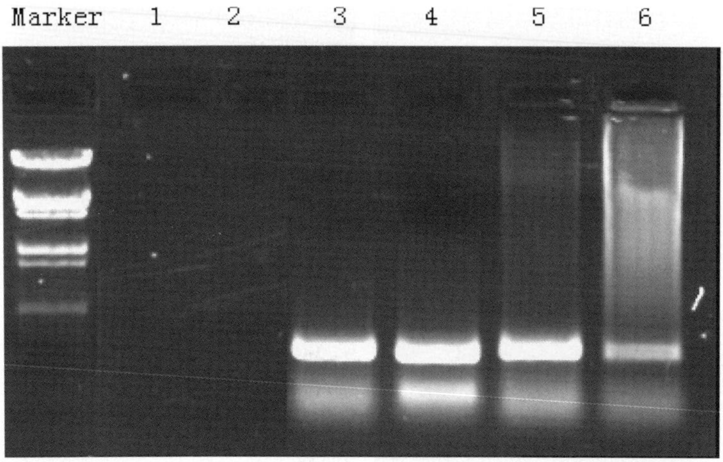


**Fig. 1.** RT-PCR detection of *C. jejuni* distending toxin expression. Lane 1 — *cdtB* (negtive control); lane 2 — *gyrA* (negative control); lane 3 — 16S RNA (genome DNA positive control); lane 4 — *gyrA* (genome DNA positive control); lane 5 — *cdtB* (*C. jejuni* culture at 37°C); lane 6 — *cdtB* (*C. jejuni* culture after −70°C refrigeration); marker — λ/*Hin*dIII + *Eco*RI marker.

in the saline group, due to the lack of nutrients, and incubation in surface water resulted in remarkable changes in cell morphology, obviously due to the low osmotic pressure.

### 2.3. *RT-PCR detection of* C. jejuni *intestinal toxin gene expression*

An amplified *cdtB* gene, using the *C. jejuni gyrA* gene as the internal parameters of control, is shown in Fig. 2. *C. jejuni* could produce toxins cdtB after incubating in the M199 medium at 37°C. The results indicate that *C. jejuni* could still express a small amount of cdtB after being frozen at −70°C.

## 3. Discussion

We aimed to simulate the state of storage of chilled and frozen food and chose 4°C and −70°C to induce *C. jejuni* into the VBNC state. The anabiotic bacteria were stained to confirm their vitality. To monitor the growth

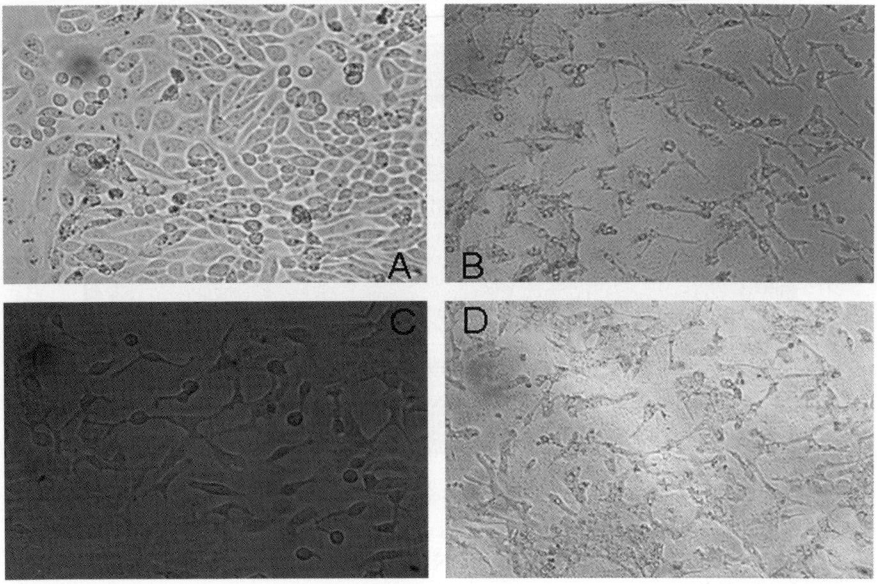


**Fig. 2.** HeLa and VeroE6 cells testing for *C. jejuni* toxicity. Detection of cytotoxic factors in culture filtrates of *C. jejuni* after 48 h culture (aliquot 50 μl in each hole). (A) Normal HeLa cells. (B) HeLa cells after adding filtrate 36 h later; elongation of cells and intracellular dark granules are shown. (C) Normal Vero cells. (D) Vero cells after adding filtrate; elongation of cells and partial lysis of cells are shown.

capability of these bacteria, the revived bacteria were further cultured to a certain concentration and stored at 4°C for two weeks (–70°C for five days). After CTC staining, the total bacterial count, the viable count, and the bacterial number which could grow on the goat blood plate, were calculated. The results showed that the count of the bacteria stored at 4°C decreased by two orders of magnitude while those stored at –70°C decreased by three orders of magnitude. Though storage at low temperature resulted in the death of a certain amount of *C. jejuni*, if the contaminated number of bacteria was high, the remnant live bacteria could still cause infections.

In this study, the culturable number of bacteria from the plate count was one order magnitude lower than that from the viable count. This showed that *C. jejuni* in the VBNC state could not revive on a general goat blood plate. Therefore, the *C. jejuni* numbers detected on the blood plate were less than the actual number of *C. jejuni* including the VBNC state in samples.

Our results also indicated that using the change in the morphology of cultured cells is feasible in detecting *C. jejuni* enterotoxin. Due to the limitation of facilities, we did not detect CE and other cytotoxins. We chose the supernatant of *C. jejuni* which has undergone low-temperature treatment to study their capability of toxin production, which may be useful for food safety surveillance. In this study, by M199 medium culture, the revived bacteria could be assayed for their toxin productivity. Since there have been a number of reports on detecting *C. jejuni* intestinal toxin gene expression at the RNA level, we used the published primer sequence to confirm the change in the *C. jejuni*'s ability to produce toxins in our experiments.

## References


1. Kist M *et al.*, in *Proceedings of the 3rd International Workshop on* Campylobacter *Infections*, ed. Pearson AD (Public Health Laboratory Service, 1985), pp. 23–27.
2. Siddique AB *et al.*, *J Trop Med Hyg*, 1991, **94**: 175–179.
3. Wu R *et al.*, *Chin J Vet Sci Technol*, 1998, **28**: 9–12.
4. Johnson WM *et al.*, *J Clin Microbiol*, 1986, **24**(2): 275–281.
5. Xu HS *et al.*, *Marine Sci*, 2000, **6**: 52–54.
6. Rollins DM *et al.*, *Appl Environ Microbiol*, 1986, **52**: 531–538.

# SECTION 2

# Viruses

# Introduction and Comments: Trends on Virus Research in China

Yu-Mei Wen

*Key Laboratory of Medical Molecular Virology, Shanghai Medical College, Institutes of Biomedical Sciences, Fudan University, Shanghai 200032, China*

In the past several decades, basic research on viruses in China has been dominated by studies on the hepatitis B virus (HBV). This is partly due to the fact that more than 10% of the population in China were hepatitis B surface antigen (HBsAg) carriers, and chronic hepatitis patients and hepatocellular carcinoma patients were infected with HBV; and partly because a large number of grants have been awarded to those who were carrying out research in this area. The preventive vaccination program on hepatitis B has resulted in huge success, by reducing the general HBsAg carrier rate to 7.18% (announced by the Chinese Ministry of Health in 2008). Nevertheless, there are still more than one hundred million Chinese people infected with HBV, and among them 30 million are patients. In this volume a paper (Chapter 2.2) on bioinformatics analysis of data from NCBI among hundreds of the envelope genes of HBV is presented, predicting the potential functional sites that merit biological studies. By comparing an HBsAg positive transgenic mouse model with its nontransgenic siblings, the transcriptomic and the proteomic analysis of the liver tissues between these animals are presented (Chapter 2.1). These papers show one of the trends in HBV research, which is to search for new targets for designing new anti-HBV drugs. Another article (Chapter 2.3) related to the hepadnavirus reports that using the duck hepatitis B animal model, one can evaluate the inactivation of blood components by the methylene blue photodynamic method, which presents a new application of this animal model.

Aside from HBV, there are a small number of studies on other viruses. Isolation of viruses from clinical samples was thought not rapid enough for clinical diagnosis, but it is crucial for understanding the evolution of viruses and for studies on the molecular epidemiology of viral infections. Virus isolation and sequencing has been carried out by the Chinese Center for Disease Control and Prevention during outbreaks of SARS, surveys of influenza and avian influenza viruses, and HIV studies. However, for other infections, isolation of viruses from clinical specimens was rare. The SH gene and its flanking region of a viral strain isolated from a mumps case represents the current status of viral genetic studies in China (Chapter 2.4).

In this volume an article (Chapter 2.5) on the HIV-related knowledge, attitude and behavior among policewomen and female prisoners is included to show that for AIDS control, not only is scientific research important, but investigations into the attitudes of society should be carried out. In fact, HIV infection is not only a medical issue but also a social one. By publishing such studies we will allow the authorities and policymakers to know in depth the true outcomes of the HIV education program for the public and for people at high risk.

All in all, virologists in China are progressing in taking the lead in studies on microbes and infection. Facing the future, we hope a large variety of viruses will be studied (with support), including vector-borne viruses, zoonotic viral diseases, and prions.

# 2.1

# Transcriptomic and Proteomic Studies on HBsAg-Positive Transgenic Mouse Livers Suggest Alterations in Carbohydrate, Lipid and Amino Acid Metabolism

Jun Ren*, Chao Zhao*, Cai-Yun Fang†, Zhang-Mei Ma*, Peng-Yuan Yang† and Yu-Mei Wen*,‡

**Key Laboratory of Medical Molecular Virology, Shanghai Medical College,*
*†Institutes of Biomedical Sciences, Fudan University, Shanghai 200032, China*

**Objective.** To study alterations in liver metabolism by persistent expression of the hepatitis B surface antigen (HBsAg) in a transgenic mouse model.

**Methods.** The Affymetrix DNA microarray was used to study differential gene expression profiles of the transgenic mouse liver versus those in the controls. Proteomic experiments were conducted to identify the alteration of metabolism-related enzymes in transgenic mouse livers. Data were processed by bioinformatics analyses.

**Results.** DNA microarray showed that 43 genes were (≥ 2 folds) upregulated and 104 genes were downregulated in the transgenic mouse

‡Corresponding author. Email: ymwen@shmu.edu.cn

livers. Eighteen metabolism-related enzymes were identified by proteomic studies with higher levels (≥ 1.5 folds) and 9 enzymes with lower levels in the transgenic mouse livers. The persistence of HBV proteins affected host hepatocellular metabolism. Specifically, cholesterol biosynthesis increased while bile acid synthesis decreased. Glycolysis was reduced, but gluconeogenesis was increased. Amino acid catabolism was increased and urea synthesis was decreased.

**Conclusion.** The potential biological impacts of these metabolic alterations in HBV persistent infections are discussed.

*Keywords*: hepatitis B virus; transgenic mouse; DNA microarray; proteomic; metabolism.

## 1. Introduction

More than 350 million people are infected with hepatitis B virus (HBV) worldwide, and one third of the cases will lead to chronic hepatitis, cirrhosis or hepatocellular carcinoma (HCC).[1] Uniquely, in the sera of the majority of hepatitis B surface antigen (HBsAg) carriers, a high titer of HBsAg could be detected though viral replication was low or even undetectable. To date, the mechanisms of how the host supported synthesis and secretion of HBsAg and the effects of HBsAg persistence on host metabolism have not been fully studied. All current anti-HBV drugs are targeted at inhibition of viral replication, while none is aimed at eliminating persistence of HBsAg. Thus, basic studies on the interactions between the persistence of HBsAg and hepatocytes will ultimately help to develop novel therapeutic strategies for clearance of HBsAg.

A lineage of transgenic mouse #59 is an animal model for persistent infection of HBV. A full-length sequenced HBV isolate (GenBank accession number AF461363) was integrated into chromosome 9 of the C57BL/6J mouse, in which the concentration of serum HBsAg was persistently higher than 100 ng/ml, while no viral particles were detected.[2] Therefore, these mice mimic HBsAg chronic carriers. DNA microarray results on liver tissues of mouse #59 were compared with those on control mouse livers to observe the alteration of gene expression profiles.

Differences in protein abundance of metabolism-related enzymes in the mouse liver were compared using the proteomic approach.[5] Alterations detected were studied using bioinformatic analysis, and results from both approaches are discussed.

## 2. Results

### 2.1. *Microarray analysis and real-time quantitative RT-PCR assay of liver tissue of the transgenic mouse*

An Affymertix DNA microarray was used to observe the differential gene expression of the liver of transgenic mouse #59. Forty-three genes were upregulated more than two-fold, while 104 genes were downregulated more than two-fold in the liver tissues of transgenic mice compared with the control. The analysis and categorization of genes were performed by Web-based analytic protocol Gene Ontology (GO) from Affymetrix (Tables 1 and 2). To confirm the results from the microarray assay, 10 genes

**Table 1.** Genes upregulated in the liver of transgenic mouse lineage #59.

| Category | Gene name | Gene symbol | Fold change |
|---|---|---|---|
| Amino acid metabolism | Serine dehydratase | Sds | 2.0 |
| Carbohydrate metabolism | Peroxisome proliferator-activated receptor alpha | Ppara | 2.0 |
| Lipid and sterol metabolism | Lipin 1 | Lpin1 | 4.0 |
| | Lipin 2 | Lpin2 | 2.0 |
| | Peroxisome proliferator-activated receptor alpha | Ppara | 2.0 |
| Other metabolism | RIKEN cDNA 1700124F02 gene | 1700124F02Rik | 5.0 |
| Cell growth, death and development | Growth arrest-specific 1 | Gas1 | 2.0 |
| | Insulin-like growth factor-binding protein 5 | Igfbp5 | 2.6 |

(*Continued*)

**Table 1.** (*Continued*)

| Category | Gene name | Gene symbol | Fold change |
|---|---|---|---|
| Cytoskeleton and extracellular matrix | C-type lectin-related f | Clrf | 2.6 |
| Immuno- and acute phase response | Leukocyte cell-derived chemotaxin 1 | Lect1 | 2.0 |
| | X-linked lymphocyte-regulated 4 | Xlr4 | 2.3 |
| Protein modification | FK506-binding protein 5 | Fkbp5 | 3.0 |
| | Sialyltransferase 9 (CMP-NeuAc:lactosylceramide alpha-2,3-sialyltransferase) | Siat9 | 3.0 |
| Signal transduction | Regulator of G-protein signaling 16 | Rgs16 | 7.0 |
| Transcription regulation | D site albumin promoter-binding protein | Dbp | 11.0 |
| | Inhibitor of DNA binding 4 | Idb4 | 3.0 |
| | Zinc finger protein 288 | Zfp288 | 2.6 |
| | Kruppel-like factor 1 (erythroid) | Klf1 | 2.0 |
| | Peroxisome proliferator-activated receptor alpha | Ppara | 2.0 |
| | Nucleoplasmin 3 | Npm3 | 2.0 |
| | Thyrotroph embryonic factor | Tef | 2.0 |
| | Similar to zinc finger protein 97 | LOC233987 | 2.0 |
| | Nuclear receptor subfamily 1, group D, member 2 | Nr1d2 | 3.5 |
| | Peroxisome proliferator-activated receptor alpha | Ppara | 2.0 |
| Transporter | Amiloride-sensitive cation channel 5, intestinal | Accn5 | 4.0 |
| | Cytochrome P450, family 17, subfamily a, polypeptide 1 | Cyp17a1 | 3.2 |
| | Cytochrome P450, family 4, subfamily a, polypeptide 14 | Cyp4a14 | 3.2 |
| | Solute carrier organic anion transporter family, member 1a4 | Slco1a4 | 2.0 |
| Ubiquitination | Ubiquitin-specific protease 2 | Usp2 | 4.6 |

(*Continued*)

**Table 1.** (*Continued*)

| Category | Gene name | Gene symbol | Fold change |
|---|---|---|---|
| Others | Expressed sequence AI604832 | AI684832 | 223.0 |
| | RIKEN cDNA 2610524G07 gene | 2610524G07Rik | 5.0 |
| | Thyroid hormone-responsive SPOT14 homolog (Rattus) | Thrsp | 5.0 |
| | Tripartite motif protein 16 | Trim16 | 4.0 |
| | Similar to promyelocytic leukemia zinc finger protein (LOC235320), mRNA | / | 3.0 |
| | RIKEN cDNA 2310032D16 gene | 2310032D16Rik | 3.0 |
| | RIKEN cDNA B230342M21 gene | B230342M21Rik | 2.6 |
| | X-linked lymphocyte-regulated 4 | Xlr4 | 2.3 |
| | Expressed sequence AU044919 | AU044919 | 2.0 |
| | RIKEN cDNA 1810015C04 gene | 1810015C04Rik | 2.0 |
| | Tissue inhibitor of metalloproteinase 3 | Timp3 | 2.0 |
| | Leukocyte cell-derived chemotaxin 1 | Lect1 | 2.0 |
| | RIKEN cDNA 0610012A05 gene | 0610012A05Rik | 2.0 |
| | RIKEN cDNA A830037N07 gene | A830037N07Rik | 2.0 |

**Table 2.** Genes downregulated in the liver of transgenic mouse lineage #59.

| Category | Gene name | Gene symbol | Fold change |
|---|---|---|---|
| Amino acid metabolism | Asparagine synthetase | Asns | −2.0 |
| Carbohydrate metabolism | Phosphofructokinase, liver, B-type | Pfkl | −2.0 |
| Lipid and sterol metabolism | Mevalonate (diphospho) decarboxylase | Mvd | −2.0 |
| | RIKEN cDNA 1300017C10 gene | 1300017C10 Rik | −2.3 |
| | Sterol-C5-desaturase (fungal ERG3, delta-5-desaturase) homolog (*S. cerevisae*) | Sc5d | −2.5 |

(*Continued*)

**Table 2.** (*Continued*)

| Category | Gene name | Gene symbol | Fold change |
|---|---|---|---|
| Other metabolism | Monooxygenase, DBH-like 1 | Moxd1 | –20.0 |
| | Phosphatidylinositol glycan, class A | Piga | –3.0 |
| | Deoxycytidine kinase | Dck | –3.0 |
| | Retinol-binding protein 1, cellular | Rbp1 | –2.0 |
| | Sepiapterin reductase | Spr | –2.0 |
| | Lipopolysaccharide-binding protein | Lbp | –2.0 |
| Cell growth, death and development | Growth arrest and DNA-damage-inducible 45 gamma | Gadd45g | –6.0 |
| | Tumor necrosis factor receptor superfamily, member 10b | Tnfrsf10b | –5.0 |
| | Bcl-associated death promoter | Bad | –3.0 |
| | Helicase, lymphoid-specific | Hells | –4.0 |
| | Insulin-like growth factor-binding protein 1 | Igfbp1 | –2.3 |
| | M phase phosphoprotein 6 | Mphosph6 | –5.0 |
| | FBJ osteosarcoma oncogene | Fos | –4.0 |
| | Early growth response 1 | Egr1 | –2.0 |
| | src homology 2 domain-containing transforming protein C1 | Shc1 | –2.0 |
| | Cytokine inducible SH2-containing protein | Cish | –2.0 |
| | Jun oncogene | Jun | –3.0 |
| Cytoskeleton and extracellular matrix | Brevican | Bcan | –19.0 |
| | Intercellular adhesion molecule | Icam1 | –2.0 |
| | Tubulin, beta 2 | Tubb2 | –2.0 |
| | Microtubule-associated protein, RP/EB family, member 2 | Mapre2 | –2.0 |
| | Profilin 2 | Pfn2 | –2.0 |
| | Diaphanous homolog 1 (*Drosophila*) | Diap1 | –2.0 |
| | RIKEN cDNA 5730592L21 gene | 5730592L21 Rik | –2.5 |
| Immuno- and acute phase response | Chemokine (C-X-C motif) ligand 14 | Cxcl14 | –11.0 |
| | Orosomucoid 2 | Orm2 | –2.5 |
| | Serum amyloid A 3 | Saa3 | –16.0 |
| | Chitinase 3-like 3 | Chi3l3 | –3.5 |

(*Continued*)

**Table 2.** (*Continued*)

| Category | Gene name | Gene symbol | Fold change |
|---|---|---|---|
| | Chemokine (C-X-C motif) ligand 9 | Cxcl9 | –2.0 |
| | Toll-like receptor 2 | Tlr2 | –3.0 |
| | S100 calcium-binding protein A8 (calgranulin A) | S100a8 | –2.0 |
| | Milk fat globule-EGF factor 8 protein | Mfge8 | –2.0 |
| | Serum amyloid A 2 | Saa2 | –2.6 |
| | B-cell translocation gene 2, antiproliferative | Btg2 | –3.0 |
| | B-cell translocation gene 2, antiproliferative | Btg2 | –3.0 |
| | Melanoma antigen, family E, 1 | Magee1 | –2.3 |

with different significance were selected for real-time PCR. Good correlation of up- or downregulated genes was shown between the microarray assay and real-time PCR (Table 3), except for one gene which showed conflicting results between PCR and microarray assays.

## 2.2. *Proteomic analysis of liver tissue of the transgenic mouse*

As gene expression profiles in the liver tissues between the transgenic mouse and the controls revealed differences in expression levels among genes involved in metabolism and growth, the proteomic approach was utilized to study the proteins with different abundances in the liver tissues in the transgenic mouse and controls. In total, 126 proteins with different abundances more than 1.5-fold were identified by MS. Some of the enzymes with different abundances related to metabolism of sugar, lipid and amino acids are analyzed and discussed (Tables 4 and 5).

# 3. Discussion

Utilizing transgenic mouse lineage #59 as an animal model to mimic the human HBV carrier, alterations of hepatocellular gene expression profiles are shown by DNA microarray, and subsequently alterations of

**Table 3.** Real-time quantitative RT-PCR assay for selected genes.

| Gene name | Gene symbol | Fold changes of differential expression* | | |
|---|---|---|---|---|
| | | DNA microarray | Real time RT-PCR | Real time RT-PCR (new batch of mice) |
| Expressed sequence AI604832 | AI604832 | 223 | 18 | 25 |
| Chemokine (C-X-C motif) ligand 14 | Cxcl14 | −11 | −4 | −2 |
| D site albumin promoter binding protein | Dbp | 11 | 31 | 6 |
| Epstein–Barr virus-induced gene 2 | Ebi2 | −5 | −7 | −5 |
| FBJ osteosarcoma oncogene | Fos | −4 | −6 | −18 |
| Insulin-like growth factor-binding protein 1 | Igfbp1 | −2 | −8 | −2 |
| Monooxygenase, DBH-like 1 | Moxd1 | −20 | −23 | −3 |
| Regulator of G-protein signaling 16 | Rgs16 | 7 | 6 | 3 |
| Serum amyloid A 3 | Saa3 | −16 | −60 | −8 |
| Tumor necrosis factor receptor superfamily, member 10b | Tnfrsf10b | −5 | −17 | 2 |

*Numbers are fold changes of differential expressed genes in the liver of the transgenic mice over the control mice. A positive value represents upregulation and a negative value represents downregulation.

enzymes with different abundances related to metabolism of carbohydrate, lipid and amino acids are observed by the proteomic approach. Although no specific identical gene or protein was discovered by the two methods used, a trend of alteration was found by both approaches. Several genes involved in the same metabolic process, especially genes of the rate-limiting enzymes, showed a consistent trend of alteration, which implied that certain metabolic processes were changed in the liver tissue of the transgenic mouse.

HBsAg is secreted as lipoprotein particles from the liver cells into the serum, in which the host-derived lipid makes up about 25% of the particle mass and cholesterol about 15% of the lipid composition. It was

**Table 4.** Partial metabolism-related enzymes with higher abundance in the liver of transgenic mouse lineage #59 over the control mice.

| Category | Gene name | Gene symbol | Fold change |
|---|---|---|---|
| Amino acid metabolism | Arginase 1, liver | Arg1 | 1.9 |
| | Branched chain ketoacid dehydrogenase E1, beta polypeptide | Bckdhb | 1.9 |
| | Glutamate dehydrogenase (*Mus musculus*) | Glud1 | 2.0 |
| | Aspartoacylase (aminoacylase) 3 | Acy3 | 1.6 |
| | Methionine adenosyltransferase I, alpha | Mat1a | 1.6 |
| | N-sulfotransferase; tyrosine-ester sulfotransferase (*Mus musculus*) | Sult1d1 | 1.7 |
| | D-dopachrome tautomerase | Ddt | 2.4 |
| | Ornithine aminotransferase | Oat | 5.1 |
| | S-adenosylhomocysteine hydrolase | Ahcy | 4.1 |
| Carbohydrate metabolism | Aflatoxin B1 aldehyde reductase 1 (*Mus musculus*) | Akr7a5 | 3.3 |
| | Aldolase 2, B isoform | Aldob | 1.6 |
| | Ketohexokinase | Khk | 1.6 |
| | Sorbitol dehydrogenase 1 | Sdh1 | 8.3 |
| | Phosphoglycerate kinase 1 | Pgk1 | 4.3 |
| | Phosphomannomutase 2 | Pmm2 | 3.8 |
| | Fructose bisphosphatase 1; FBPase liver (*Mus musculus*) | Fbp1 | 2.6 |
| Lipid and sterol metabolism | Farnesyl diphosphate synthetase | Fdps | 2.9 |
| | Crystallin, lambda 1 | Cryl1 | 2.4 |

speculated that the persistence of abundant HBsAg in sera of HBV patients would affect the synthesis of lipid in host liver cells. Lin *et al.* have shown that diminished cholesterol synthesis inhibited HBsAg secretion in Hep3B and HepG2.2.2.15 cells.[3] In this study, some enzymes in biosynthesis of cholesterol, such as mevalonate decarboxylase (MVD), sterol-C5-desaturase (SC5D) and farnesyl diphosphate synthase (FDPS), were found changed in the mRNA level or the protein level in the HBV transgenic mouse compared to the controls, which suggested the enhancement of biosynthesis of cholesterol. On the other hand, aldo-keto reductase family 1, member D1 (AKR1D1), the rate-limiting enzyme in bile acid

**Table 5.** Partial metabolism-related enzymes with lower abundance in the liver of transgenic mouse lineage #59 over the control mice.

| Category | Gene name | Gene symbol | Fold change |
|---|---|---|---|
| Amino acid metabolism | Glycine N-methyltransferase | Gnmt | −2.0 |
| | Ornithine transcarbamylase | Otc | −2.6 |
| Carbohydrate metabolism | Aldo-keto reductase family 1, member C14 | 9030611N15Rik | −2.6 |
| | Glycerol phosphate dehydrogenase 1, cytoplasmic adult (*Mus musculus*) | Gpd1 | −1.8 |
| | Malate dehydrogenase, soluble (*Mus musculus*) | Mdh1 | −1.6 |
| | Galactokinase (*Mus musculus*) | | −2.3 |
| Lipid and sterol metabolism | Paraoxonase 1 | Pon1 | −2.3 |
| | Triosephosphate isomerase (*Mus musculus*) | Tpi | −2.1 |
| | Aldo-keto reductase family 1, member D1 | Akr1d1 | −3.0 |

synthesis, was decreased in the protein level. This finding suggested that biosynthesis of bile acid was reduced; it is involved in the consumption of cholesterol. It is known that endogenous cholesterol exerts negative feedback activity on the rate-limiting enzyme of cholesterol biosynthesis, HMGCoA reductase (HMGCR).[6] However, no detectable alteration of HMGCR expression was observed in our experiments, suggesting that there was no accumulation of intracellular cholesterol, and the dynamic metabolism of intracellular cholesterol seemed not impaired. Taken together, *in vivo* expression of HBsAg seemed to cause upregulation of host cholesterol synthesis and downregulation of bile acid synthesis by altering certain pathways in the metabolic processes. One may speculate that, by increasing cholesterol synthesis together with persistent secretion of HBsAg particles into the serum to dispose of the increased cholesterol, the overall intracellular cholesterol level might be kept balanced. This speculation seemed to be consistent with clinical observation: the serum cholesterol level was within the normal limit in HBV-infected patients.

Phosphofructokinase (PFK1) is one of the rate-limiting enzymes in glycolysis. The decrease of PFK1 mRNA in the liver of the HBV transgenic mouse suggested that there was inhibition of host glycolysis. Although the protein level of PFK1 was not found to be changed by the proteomic approach, the upregulation of ALDOB, ketohexokinase (KHK) and PGK1 in our experiments showed that there was decrease in host glycolysis. On the other hand, FBP1, the key enzyme in gluconeogenesis — the reverse of glycolysis — was found in high abundance in the liver of the HBV transgenic mouse, which implied increased host gluconeogenesis. The possibility of sugar accumulation caused by the decreased glycolysis and increased gluconeogenesis deserves to be studied. Toshkov *et al.* described "glycogen-storage foci" during hepatocarcinogenesis in HBV transgenic mice.[4] Whether this phenomenon of glycogen accumulation correlates with our current observation is to be determined.

A potential problem caused by the inhibition of glycolysis is lack of energy. In our study, the catabolism of amino acids was found upregulated, which might have been to compensate for energy supply. Ornithine transcarbamylase (OTC) in the liver of the HBV transgenic mouse was reduced to 38% of the control, suggesting a lowered level of urea synthesis. Urea synthesis is an important removing pathway for free ammonia produced in amino acid catabolism. High catabolism of amino acids and low synthesis of urea would lead to accumulation of free ammonia in tissues and serum, which could result in dysfunction of the liver and brain.[7]

The advantage of using the transcriptomic and proteomic approaches to compare gene expression profiles and abundance of proteins in the liver between the current HBsAg transgenic mouse and its counterparts is that the transgenic mice and the control mice were generated in the same laboratory, under the same breeding conditions with the same genetic background. Thus, results were highly comparable. More lineages of transgenic mice should be employed for study to exclude the impact of HBV integrated sites in the host cellular chromosome. Besides, pooled samples from several mice could miss the alterations of the proteins in low abundance though individual diversity might be avoided by pooled samples. Furthermore, the sensitivities of the approaches might influence the results.

In summary, the majority of the genes up- or downregulated and the altered abundance among proteins were metabolism-related enzymes, which suggested that persistence of HBsAg could affect the metabolic pathways in hosts. From the technical point of view, since the transcriptomic approach was more sensitive than the proteomic approach, to explore in comparative studies, one should begin with the transcriptomic approach to find the potential targets, followed by the proteomic approach, or simply study changes at the protein level, to confirm the findings.

## References

1. Ganem D *et al.*, *N Engl J Med*, 2004, **350**: 1118–1129.
2. Lin X *et al.*, *J Med Virol*, 2001, **64**: 299–304.
3. Lin YL *et al.*, *Virology*, 2003, **314**: 253–260.
4. Toshkov I *et al.*, *Hepatology*, 1994, **20**: 1162–1172.
5. Zhao C *et al.*, *J Med Virol*, 2007, **79**: 1478–1484.
6. Meigs TE *et al.*, *J Biol Chem*, 1996, **271**: 7916–7922.
7. Tuchman M *et al.*, *J Med Genet*, 1995, **32**: 680–688.

# 2.2

# Large-Scale Polymorphism Analysis of Hepatitis B Virus Envelope Protein Using Bioinformatics

Zhe Chen*, Yue Zhu†, Yi-Xue Li† and Yu-Mei Wen*,‡

**Key Laboratory of Medical Molecular Virology, Shanghai Medical School, Institutes of Biomedical Sciences, Fudan University, Shanghai 200032, China*

*†Shanghai Center for Bioinformation Technology, Shanghai 200235, China*

**Objective.** Using the Bio-HBV database, the polymorphism of HBV envelope protein (HBsAg, pre-S) was studied using bioinformatics.

**Methods.** The Shannon entropy value was used to analyze the conservation of each amino acid in the sequence of this protein. With the BLOSUM90 scoring matrix and PAML (phylogenetic analyses by maximum likelihood) software package, abnormal amino acid substitution patterns, which reflected specific selective pressure, were searched for and identified.

**Results.** Within these substitution patterns, ps35, ps132, ps49 and s127 substitutions were found to have high statistical significance. Multiple functional regions located within the envelope protein were also analyzed using bioinformatics methods.

‡Corresponding author. Email: ymwen@shmu.edu.cn

**Conclusion.** Results revealed that there was a close association between the conserved amino acid sequences and the described functional domains. In addition, based on structural analysis, several domains were predicted to have potential functions, which could be pursued for their biological implications.

*Keywords*: hepatitis B virus; envelope protein; polymorphism; pre-S1 region; pre-S2 region; S region.

## 1. Introduction

The hepatitis B virus (HBV) causes acute and chronic hepatitis, which severely threaten the health of approximately 350 million people worldwide. Among these patients, half are currently living in Asia, particularly China.[1] HBV is a DNA virus consisting of an outer lipid envelope protein and an inner icosahedral nucleocapsid core protein. Envelope proteins play a key role in inducing immunological responses and initiating host invasion. In order to find new functional domains or sites with biological significance, a Bio-HBV sequence database with HBV sequences was first constructed, and large-scale bioinformatics analysis of the sequences of HBV envelope proteins was subsequently performed.

HBV envelope proteins consist of L, M and S glycoproteins, all of which share the same open reading frame (ORF). L (pre-S1, pre-S2 and S coding frame), M (pre-S2 and S coding frame) and S (S coding frame) proteins are distinguished by different initiation codons of translation. The pre-S1 and pre-S2 regions encoding L or M proteins are recognized as pre-S regions.

Envelope proteins are associated with multiple biological functions, including viral envelopment,[2–5] adhesion[6] and infectivity.[7] Furthermore, immunogenicity of envelope proteins, including multiple B cell epitopes that have neutralization functions, has been reported.[8–10] Specifically, certain missense mutations in the "a" determinant (a B cell epitope) have been widely studied. These mutations have been shown to cause a false negative result of detecting the hepatitis B surface antigen (HBsAg) using conventional antibodies rather than the HBsAg themselves, and have also failed to induce active or passive immunity.[11,12] Glycosylation of envelope proteins is also important for the local conformation of epitopes or virion secretion.[13] The 2D topological structure of envelope proteins can be predicted using bioinformatics methods. Generally, the S region contains four transmembrane fragments, while the pre-S1 and pre-S2 regions are hydrophilic and are located outside the virion.[14] Chen *et al.*[12] reported a 3D structure of the major hydrophilic region (MHR) of S protein using a filamentous phage peptide library, which may help further in the study of the relationship between mutations and structure of the MHR.

## 2. Results

### 2.1. *Comparison between the conservation of the S region and of the pre-S region*

To exclude the influence of different genotypes, 50 complete S region sequences from each genotype (A-F genotypes) were randomly obtained through the Bio-HBV database. Twenty complete pre-S region sequences for each genotype were obtained, as stated above. Conservation of the S and pre-S regions was compared by (1) the synonymous/nonsynonymous rate at the nucleic acid level, and (2) the entropy value at the amino acid level (Table 1). Results from this analysis showed that the overall conservation of the S region was higher than that of the pre-S region and that this was statistically significant

**Table 1.** Sequence (DNA/protein) conservation in the S and pre-S regions.

| Region | Amino acid sequence[1] | | DNA sequence[2] | |
|---|---|---|---|---|
| | Entropy ($Hx \pm$ ED) | $dN$ [a] | $dS$ [b] | $\omega$ ($dN/dS$)[c] |
| S region | 0.150 ± 0.237 | 0.042 | 0.091 | 0.448 |
| Pre-S region | 0.297 ± 0.362 | 0.090 | 0.317 | 0.234 |
| Pre-S1 region | 0.294 ± 0.382 | 0.089 | 0.358 | 0.195 |
| Pre-S2 region | 0.305 ± 0.317 | 0.093 | 0.242 | 0.339 |

[1] The Shannon entropy value was used for evaluating the conservation of protein sequences. A higher entropy value represents lower conservation. $P_{\text{S, pre-S}} < 0.01$, $P_{\text{pre-S1, pre-S2}} > 0.05$. [2] By dnasp 4.0, the nonsynonymous/synonymous mutation ratio was analyzed in aligned DNA sequences. [a] $dN$ — nonsynonymous mutation ratio; [b] $dS$ — synonymous mutation ratio; [c] $\omega$ — $dN/dS$.

($P < 0.001$), which suggested that higher selective pressure might have occurred at the extracellular pre-S region.

## 2.2. *Conservation and substitution patterns of amino acids in the pre-S (pre-S1 and pre-S2) regions*

Entropy values for each site in the pre-S region are plotted in Fig. 1 and are shown to be highly variable. Furthermore, as shown in Fig. 2, there are abnormal amino acid substitution patterns at some sites with significant physiochemical changes.

In the pre-S region, the positive amino acids under selective pressure were discovered using the PAML software package, and included ps35, ps132 and ps160 (Table 2). As previously mentioned, these sites coincide with the spacer domain of HBV polymerase protein. Because the spacer domain was

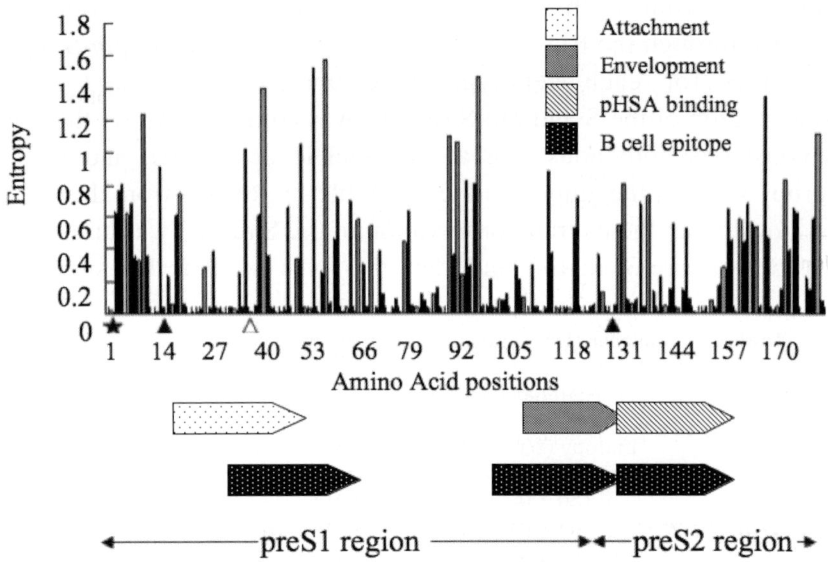


**Fig. 1.** Conservation of each amino acid position in the pre-S (pre-S1/pre-S2) region. The *X* axis represents each amino acid position (174 amino acids in total). The *Y* axis represents the corresponding entropy value. Known functional domains or immune epitopes are listed below. Filled triangle — known *N*-glycosylation positions; empty triangle — potential *N*-glycosylation position; pentagram — known myristylation position.

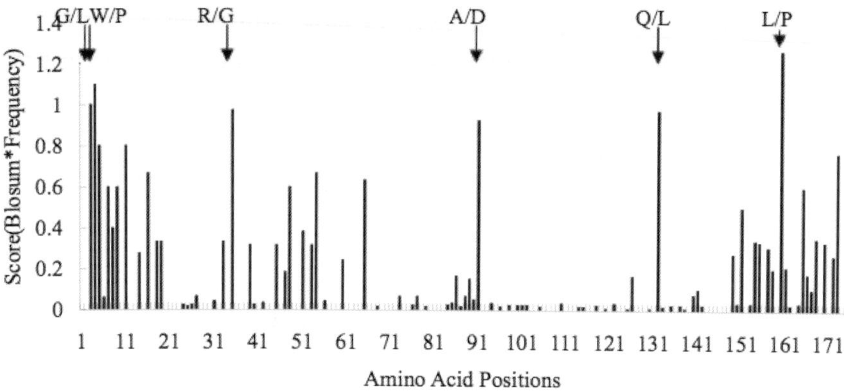


**Fig. 2.** Abnormal amino acid substitution patterns in the pre-S region. The *X* axis represents each amino acid position. The *Y* axis represents the corresponding abnormal amino acid substitution scores. ps160, ps4, ps3, ps35, ps132 and ps91 with substitution patterns L-P, W-P, G-L, R-G, Q-L and A-D, respectively, represent the most abnormal substitution patterns.

**Table 2.** Positive selective positions in the pre-S region using BEB methods.

| Position[1] | Posterior probability[2] | $\omega$ (*dN*/*dS*)[3] | Amino acid substitution pattern |
|---|---|---|---|
| ps35 | 0.996 | 5.816 | R-G |
| ps132 | 0.999 | 5.299 | Q-L |
| ps160 | 1.000 | 6.010 | L-P |

[1] Using the BEB method, positive selective positions, which reflected strong selective pressure, were searched by the M8 model implemented in the PAML software package. [2] Positions with posterior probability > 90% in the pre-S region. [3] Heterogeneous $\omega$ ratios among sites.[21]

reported to be negligible for polymerase function,[15] these mutations might occur because of exclusive selective pressures for the pre-S region.

## 2.3. *Bioinformatics analysis of known biological domains in the pre-S region*

Several domains in the pre-S region, which have been published with biological functions, were studied. The ps21–47 domain is involved in the adherence of the virus to the host cell, which is the first step of the infectious cycle.[6] In Fig. 1, ps35 is shown to be highly variable, while most of the other amino acids in ps21–47 are invariable. Furthermore, the substitution pattern for ps35 includes a substitution of arginine for glycine, which may change the physiochemical properties of the amino acid and the conformation of the local peptide. This result suggested that discontinuous epitopes or specific conformations, but not linear epitopes, may play important roles in the recognition step.

The ps107–119 domain is involved in virion envelopment.[2–5] Some sites in ps107–119 are also shown to be variable in Fig. 1. However, the substitution patterns for these sites were not significant according to the BLOSUM90 score, as shown in Fig. 2. This suggested that stable conformation of this domain may be important for virion envelopment.

The entropy values for the known neutralizing epitopes, ps34–59, ps95–109 and ps120–145,[8–10] were 0.397, 0.172 and 0.212 ($H_{ps34-59} > H_{ps120-145} > H_{ps95-109}$), respectively, which suggested the extent of corresponding immunological pressures on these epitopes. As shown in Fig. 2, substitution patterns in epitope ps34–59 (R35G) and epitope ps120–145 (Q132L) may cause significant changes in the physiochemical properties of the wild type virus.

Myristylation of glycine at the second amino acid of the pre-S1 region is important for viral infectivity.[7] In Figs. 1 and 2, glycine is shown to be highly conserved. Interestingly, the following ps3–12 is quite variable, with entropy values from 0.353 to 1.244. Particularly, according to the BLOSUM90 scores, W4P and G3L substitutions may change the local conformation, which may influence myristylation reaction.

### 2.4. *Topology of the S region*

Through the use of the TMHMM v2.1 and TMMOD software, the S region was predicted to contain four transmembrane fragments (s8–29, s79–100, s160–184, s189–210), which is in accordance with previous research.[14] The 2D topological structure is shown in Fig. 3.

### 2.5. *Conservation and substitution patterns of amino acids in the S region*

As shown in Table 1, the S region is more conserved than the pre-S region. However, several sites, such as s49 and s127, are highly variable, with

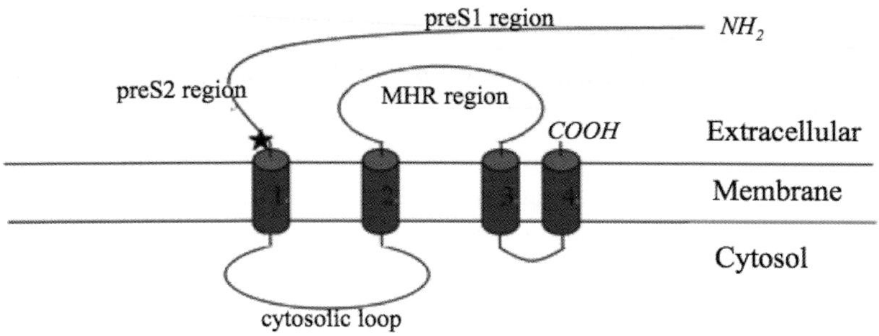


**Fig. 3.** 2-D topological model of envelope protein. s8–28, s79–100, s160–184 and s189–210 are transmembrane fragments in S protein. The pre-S1 and pre-S2 regions are completely hydrophilic, and exposed to the outer environment. Pentagram: starting point of the S region.

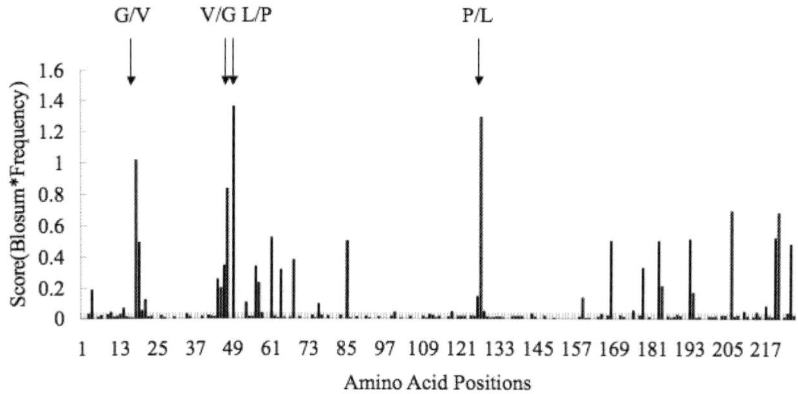


**Fig. 4.** Abnormal amino acid substitution patterns in the S region. The *X* axis represents each amino acid position. The *Y* axis represents the corresponding abnormal amino acid substitution scores. s49, s127, s18 and s47 with substitution patterns L-P, P-L, G-V and V-G, respectively, represent the most abnormal substitution patterns.

significant amino acid substitution patterns, as shown in Fig. 4 and Table 3. Furthermore, s49 coincides with rt57 and rt58 of the HBV polymerase reverse transcriptase domain, while s127 coincides with rt135 and rt136. However, the substitution patterns of corresponding sites in HBV polymerase were not changed significantly according to BLOSUM90 scores (data not shown here), which suggested that the s49 and s127 mutations occurred as a result of exclusive pressure for the S region, but not for polymerase.

**Table 3.** Positive selective positions in the S region using BEB methods.

| Position[1] | Posterior probability | $\omega$ $(dN/dS)^2$ | Amino acid substitution pattern |
|---|---|---|---|
| s49 | 0.992 | 5.223 | L-P |
| s127 | 1.000 | 5.980 | P-L |

[1] Positions with posterior probability > 90% in the S region. [2] Heterogeneous $\omega$ ratios among sites.[21]

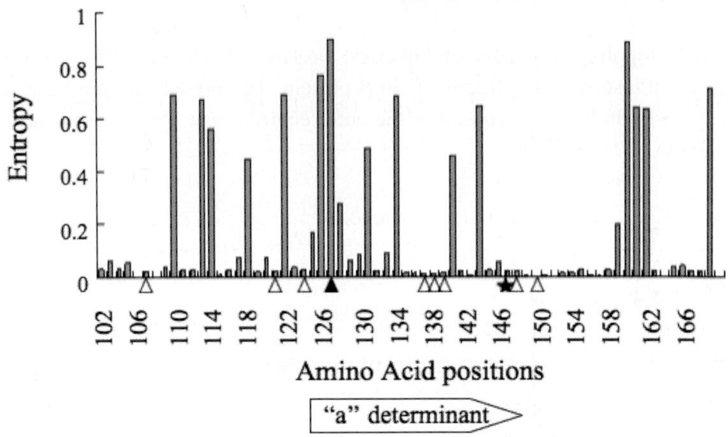


**Fig. 5.** Conservation of each amino acid position in the MHR region. The *X* axis represents each amino acid position (s100–169). The *Y* axis represents the corresponding entropy value. The "a" determinant position is listed below. Filled triangle — S127, known position associated with serotype typing; empty triangle — cysteine positions; pentagram — known *N*-glycosylation position.

## 2.6. *Bioinformatics analysis of known biological domains in the S region*

The MHR is located between the second and the third transmembrane fragment, facing the outer host environment. There is an important neutralizing epitope located inside the MHR, called the "a" determinant, whose conformation is maintained by disulfide bonds.[12,16] In Fig. 5, all the cysteines in the MHR are shown to be highly conserved according to entropy values ($Hx \ll 0.2$), which further emphasizes the importance of disulfide bonds in maintaining the conformation.

The "a" determinant was used to differentiate different serotypes of HBV.[17] As expected, the sero-typing-associated amino acids (s122, s160, s125, s126, s127) were highly variable ($Hx > 0.2$). The substitution patterns for s122 and s160 were K122R or K160R, which occurred between two alkaline amino acids with similar physiochemical properties. However, the substitution pattern for s127 is from proline to leucine, threonine or isoleucine, which may change the local conformation due to the disruption of the rigid folding of proline (Table 4). G145R mutation in the second loop of the "a" determinant has been reported for its possible role in failure of protection after vaccination with the wild type HBsAg vaccine.[11,12] However, results from the present bioinformatics study indicate that the adjacent amino acids (s140, s143) were more variable than s145 according to entropy values. This phenomenon deserves further study in order to determine the significance of the s140 and s143 mutations, especially for their possible influence on the outcome of passive immunity.

The envelopment-associated cytosolic loop of the S region is located between the first and the second transmembrane fragment. In Fig. 4, s49 is shown to be variable with substitution patterns of L49P, which may change the local conformation. Although only $S_{56-80}$ was directly involved in the binding of HBV core protein with envelopment proteins,[2–5] s49 may influence this binding step through changes of local conformation.

**Table 4.** Conservation of amino acids associated with serotype typing.

| Position[1] | Serotype classification | Entropy ($Hx$) | Major patterns[2] | Minor patterns[3] |
|---|---|---|---|---|
| S122 | d/y | 0.689 ($H > 0.2$) | K/R | — |
| S160 | r/w | 0.642 ($H > 0.2$) | K/R | — |
| S125 | w1/w2/w3/w4 | 0.169 ($H \approx 0.2$) | T/M | — |
| S126 | | 0.764 ($H > 0.2$) | T/I | T/A, T/S, T/N |
| S127 | | 0.896 ($H > 0.2$) | P/L | P/T, P/I |

[1] Known positions for serotype classification. [2] Major amino acid substitution patterns. [3] Minor amino acid substitution patterns (s122 and s160 have no minor substitutions patterns).

### 2.7. *Bioinformatics analysis of N-glycosylations in envelopment proteins*

The *N*-glycosylation reaction is important, not only for the conformation maintenance of envelopment proteins, but also for virion secretion. Potential glycosylated sites were searched according to the motif *N*-$X_1$(non-P)-*S*/*T*-$X_2$(non-P).[18,19] ps15, ps123 and s145 were shown to be conserved (*Hx*: 0.036, 0,0.020, respectively) and were determined to be *N*-glycosylation sites, as reported by previous studies.[13] Furthermore, ps37 in the pre-S1 region and s3 in the S region are potential candidates for *N*-glycosylation, as shown in Fig. 1.

## 3. Discussion

With a Bio-HBV sequence database, large-scale polymorphism analysis by bioinformatics may promote the structural and functional study of HBV proteins. Together with traditional methods, bioinformatics can not only demonstrate the known domains or sites with biological function(s), but also predict potential functional domains for further study.

In this study, the following observations demonstrate the need for further study:

(1) A ps35 mutation (R35G), which is located in an adhesive epitope,[6] or an overlapping neutralizing epitope[8] has not been studied for its function as yet. Since arginine substituted for glycine may change the conformation because of significantly different physiochemical properties, the recognition of the adhesive epitope between virion and host cells may depend upon the conformation of this epitope. In addition, a ps35 mutation may also occur because of immune escape, due to the fact that this residue is located in an overlapping neutralizing epitope.

(2) A ps132 mutation, which is also located in a neutralizing epitope,[10] and an overlapping polymerized human serum albumin (pHSA) binding site[20] may also occur. This deserves further study for its role in binding to pHSA.

(3) The ps107–109 domain, which is important for virion envelopment,[2–5] is relatively conserved, suggesting it as a possible drug target that may interfere with a critical step in the life cycle of HBV.

(4) ps3–12, which is close to the myristylation site,[7] is highly variable, with aberrant substitution patterns. In addition, ps4 was substituted by proline, which may change the local conformation. Contradictorily, the myristylation step requires a strict conformation of the local peptide, which seems inconsistent with the variability of the ps3–12 peptide and may be a starting point for further study on the importance of myristylation.

(5) s49, which is located in the cytosolic loop and near the binding site of the HBV core protein (s56–80),[2–5] was highly variable, with leucine substituted for proline. Whether this substitution will have an impact on virion envelopment deserves further study.

(6) Two potential sites for *N*-glycosylation (ps37 and s3) were discovered. Mutations in these sites may interfere with *N*-glycosylations of other sites, and may allow the virus to escape host immune responses. For example, Lee *et al.*[13] reported that simultaneous glycosylations of ps15 and ps123 would interfere with B cell immune responses.

Although some clues to potential biological functions were presented using bioinformatics, the sequences deposited in the Bio-HBV database lacked clinical background information. Furthermore, most of the sequences belonged to genotypes A, B, C and D. It is still impossible, at the present time, to perform large-scale analysis of different genotypes, including E, F, G and H. Recombination between different genotypes has also been reported.[17] On the other hand, aside from the MHR, there are still few publications regarding the complete structure of envelope proteins. Results from this study indicate that the use of new bioinformatics methods is promising for the prediction of mutations' sites and of potential biological functions associated with these mutations.

## References

1. Chisari FV *et al.*, *Annu Rev Immunol*, 1995, **13**: 29–60.
2. Loffler-Mary H *et al.*, *Virology*, 2000, **270**(2): 358–367.
3. Poisson F *et al.*, *Virology*, 1997, **228**(1): 115–120.
4. Bruss V *et al.*, *EMBO J*, 1994, **13**(10): 2273–2279.
5. Jenna S *et al.*, *Virology*, 1998, **251**(1): 176–186.

6. Neurath AR *et al.*, *Cell*, 1986, **46**(3): 429–438.
7. Bruss V *et al.*, *Virology*, 1996, **218**(2): 396–399.
8. Hu WG *et al.*, *World J Gastroenterol*, 2005, **11**(14): 2088–2094.
9. Xin ZT *et al.*, *Acta Acad Med Sin*, 2003, **25**(1): 56.
10. Leclerc C *et al.*, *Mol Immunol*, 1993, **30**(17): 1561–1572.
11. Maillard P *et al.*, *Res Virol*, 1998, **149**(3): 153–161.
12. Chen YC *et al.*, *Proc Natl Acad Sci USA*, 1996, **93**(5): 1997–2001.
13. Lee J *et al.*, *Biochem Biophys Res Commun*, 2003, **303**(2): 427–432.
14. Stirk HJ *et al.*, *Intervirology*, 1992, **33**(3): 148–158.
15. Kim Y *et al.*, *Biochem Mol Biol Int*, 1999, **47**(2): 301.
16. Mangold CM *et al.*, *Virology*, 1995, **211**(2): 535–543.
17. Bartholomeusz A *et al.*, *Rev Med Virol*, 2004, **14**(1): 3–16.
18. Gavel Y *et al.*, *Protein Eng*, 1990, **3**(5): 433–442.
19. Bause E *et al.*, *Biochem J*, 1983, **209**(2): 331–336.
20. Park JH *et al.*, *J Viral Hepat*, 2003, **10**(1): 70–79.
21. Yang Z *et al.*, *Genetics*, 2000, **155**(1): 431.

# 2.3

# Studies on the Inactivation of Hepadnavirus in Blood Components by Using a Duck Hepatitis B Virus–Infected Animal Model

Wen-Yi Wang*, Yu-Wen Huang†, Ping Lu†, Jian-Er Long*, Qin Mo†, Bing Ling†, Xiao-Ling Han*, Sheng-Fu Dong*, Di Qu*,‡ and Kai-Cheng Qian†,‡

**Key Laboratory of Medical Molecular Virology, Shanghai Medical College, Institutes of Biomedical Sciences, Fudan University, Shanghai 200032, China*
*†Shanghai Blood Center, Shanghai 200051, China*

**Objective.** To evaluate the inactivation of DNA virus by a methylene blue photodynamic method using a duck hepatitis B virus (DHBV)-infected animal model.

**Methods.** DHBV was purified by ultracentrifugation from sera. Serial diluted DHBV was added to human plasma or red blood cells (RBCs) separately, followed by inactivation using the methylene blue photodynamic method, and one-day-old ducklings were inoculated by intravenous injection. Sera from ducks were collected at intervals and detected by dot hybridization with a DHBV probe for DHBV

‡Corresponding authors. Email: dqu@shmu.edu.cn; kcqian53@sh163b.sta.net.cn

DNA, and confirmed by Southern blot hybridization for total DHBV DNA in experimental duck livers. Dot hybridization was used to compare the infectivity of DHBV ($ID_{50}$) in the blood components before and after the treatment by the methylene blue photodynamic method.

**Results.** Before inactivation, the $ID_{50}$ of DHBV–human plasma in one-day-old ducklings was $1 \times 10^{3.33}$ copies, and the $ID_{50}$ of DHBV–human RBCs in the one-day-old ducklings was $1 \times 10^{3.50}$ copies. After inactivation by the methylene blue photodynamic method, the $ID_{50}$ of DHBV–human plasma in the one-day-old ducklings was $1 \times 10^{10}$ copies, while the $ID_{50}$ of DHBV–human RBCs in the one-day-old ducklings was $1 \times 10^{8.35}$ copies. Treatment by the methylene blue photodynamic method decreased the viral titer of DHBV–human plasma by over six logs, and DHBV–human RBCs by about five logs.

**Conclusion.** Treatment by the methylene blue photodynamic method inactivates DHBV, a DNA virus, in human plasma or RBCs, and the efficacy of this procedure is more potent for the virus in plasma than that in RBCs. The DHBV-infected ducklings can be used as an *in vivo* animal model for evaluating the efficacy of DNA virus inactivation.

*Keywords*: DHBV-infected animal model; virus photodynamic inactivation.

## 1. Introduction

Many viruses, such as hepatitis B virus (HBV), hepatitis C virus (HCV) and human immunodeficiency virus (HIV), can be transmitted via blood transfusion. Posttransfusion viral diseases pose a major risk for the recipients of blood transfusion.[1,2] Although the technologies and methodologies of virus testing for blood donor selection have been improved in both sensitivity and specificity, there are still limitations on screening the virally infected donors efficiently, especially for those in the window period of viral infection at the time of donation. For example, a donor may transmit HCV before serological conversion. Therefore, posttransfusion viral infection cannot be eliminated relying merely on screening the donors by existing tests and the risk of infectious disease transmission still exists. One of the effective measures to prevent the posttransfusion viral diseases is to inactivate viruses in blood products for clinical application.[3] Thus, techniques for inactivation of viruses in plasma and other blood components have been developed. From the early 1990s, a number of laboratories have carried out investigations using photodynamic methods to inactivate viruses in plasma by photoactive compounds and visible light. The Shanghai Blood Center has established a photodynamic method for inactivating viruses in single bag plasma treated with methylene blue (MB) and visible light, and validated its efficacy in inactivation of vesicular stomatitis virus (VSV), Sindbis virus and HIV in infected cell models.[4,5] In the present study, we evaluated the efficacy of inactivation of DNA virus in blood components by MB photodynamic treatment using duck hepatitis B virus (DHBV)–infected ducks as an animal model.

## 2. Results

### 2.1. *DHBV–infected model and determination of a 50% infective dose of DHBV*

Ducklings without congenital DHBV infection were inoculated via the shank vein with $3 \times 10^8$ copies of DHBV (QP strain, $3 \times 10^9$ copies/ml in the serum). Seven days after the virus inoculation, DHBV was detected in the peripheral blood, up to nine weeks, and DHBV replication intermediates (SC, SS and CCC DNA) were detected in liver tissues. No effect of human plasma or RBC inoculation alone on the infectivity of DHBV in the ducklings was observed (data not shown).

For determination of DHBV $ID_{50}$ in the serum, DHBV at $1 \times 10^8$, $1 \times 10^6$, $1 \times 10^4$, $1 \times 0^3$ and $1 \times 10^2$ copies was separately added to human plasma or RBCs, and inoculated the ducklings. After the virus inoculation, blood was collected from each duckling at two-week intervals, for six weeks. DHBV in the inoculated duck serum was detected by dot hybridization (Fig. 1). In DHBV–human plasma–inoculated ducklings, one week after the virus inoculation DHBV viremia appeared in all the ducklings (4/4) inoculated with $1 \times 10^8$ copies, 3/4 of the ducklings inoculated with

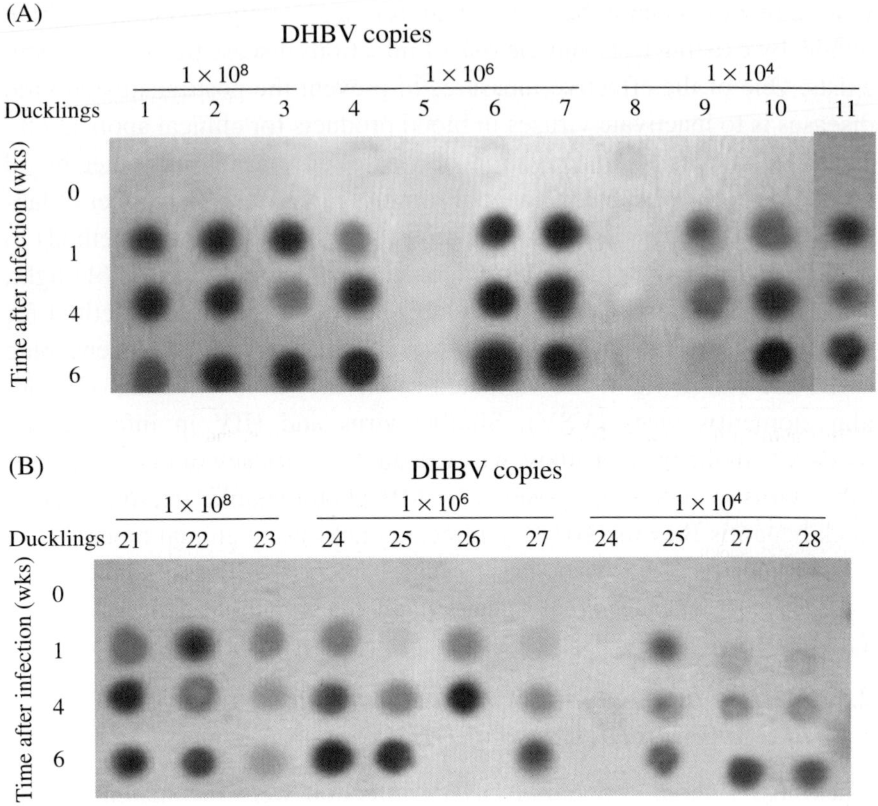


**Fig. 1.** Detection of DHBV DNA in the sera from ducklings inoculated with noninactivated DHBV in human plasma or human RBCs. One-day-old ducklings were intravenously infected with DHBV in different copies added to human plasma (DHBV-plasma) (A) or human red blood cells (DHBV-RBCs) (B). Sera were collected at intervals, and detected by dot hybridization with a DHBV DNA probe labeled with $^{32}$P. The sera from ducklings infected with $1 \times 10^3$ and $1 \times 10^2$ copies of DHBV were all negative. Therefore, the data are not shown.

$1 \times 10^6$ or $1 \times 10^4$ copies showed viremia, and no viremia was detected in the ducklings inoculated with $1 \times 10^3$ or $1 \times 10^2$ copies (Table 1). According to the results, $ID_{50}$ of DHBV–human plasma was calculated as $1 \times 10^{3.33}$ copies determined by the Reed–Muench method. The $ID_{50}$ of DHBV–human RBCs was $1 \times 10^{3.50}$. In the group of ducklings inoculated with $1 \times 10^6$ copies, one duckling showed viremia one week later and another had a seroconversion at six weeks.

**Table 1.** Determination of $ID_{50}$ for noninactivated DHBV-human plasma or DHBV–human RBCs.

| Inoculated DHBV copies[a] | Number of DHBV DNA conversions/total number[b] | | | | | |
|---|---|---|---|---|---|---|
| | DHBV–human plasma | | | DHBV–human RBCs | | |
| | 1 wk | 4 wk | 6 wk | 1 wk | 4 wk | 6 wk |
| $1 \times 10^8$ | 3/3 | 3/3 | 3/3 | 3/3 | 3/3 | 3/3 |
| $1 \times 10^6$ | 3/4 | 3/4 | 3/4 | 3/4 | 4/4 | 3/4 |
| $1 \times 10^4$ | 3/4 | 3/4 | 3/4 | 3/4 | 3/4 | 3/4 |
| $1 \times 10^3$ | 0/4 | 0/4 | 0/4 | 0/4 | 0/4 | 0/4 |
| $1 \times 10^2$ | 0/4 | 0/4 | 0/4 | 0/4 | 0/4 | 0/4 |

[a] DHBV purified from sera of DHBV-infected ducks by ultracentrifuge, and DHBV genome copy numbers were decided by dot hybridization with a DHBV DNA probe labeled with $^{32}P$. Purified DHBV was diluted as indicated in the table and added to human plasma (DHBV–plasma) or human red blood cells (DHBV–RBCs).

[b] One-day-old ducklings were intravenously infected with DHBV. Sera were collected at intervals, and detected by dot hybridization. Using the Reed–Muench method, $ID_{50}$ of DHBV-human plasma was calculated as $1 \times 10^{3.33}$, and $ID_{50}$ of DHBV–human RBCs as $1 \times 10^{3.50}$.

## 2.2. *Inactivation of DHBV infectivity by the MB photodynamic method*

Human plasma or human RBCs were incubated with $1 \times 10^{10}$ and $1 \times 10^8$ copies of DHBV separately, and treated with MB at 1 μmol/L and exposed to light (intensity > 30,000 luxes for 30 min for DHBV–human plasma, with 5 μmol/L MB, and exposure to the light for 60 min for DHBV–human RBCs). Ducklings were injected with the treated components via shank vein inoculation. After the virus inoculation, blood was collected from each duckling at two-week intervals, and DHBV was detected by dot hybridization. In the MB-treated DHBV–human plasma group, only two ducklings inoculated with $1 \times 10^{10}$ copies showed viremia and none of the ducklings inoculated with $1 \times 10^8$ copies did so (Fig. 2). Thus, $ID_{50}$ of MB-treated DHBV–human plasma was $1 \times 10^{10}$ copies. Compared with $ID_{50}$ of untreated DHBV–human plasma, MB treatment decreased the virus titer in human plasma by over six logs.

In the MB-treated DHBV–human RBCs group, all four ducklings inoculated with $1 \times 10^{10}$ copies showed viremia and 6/13 of the ducklings

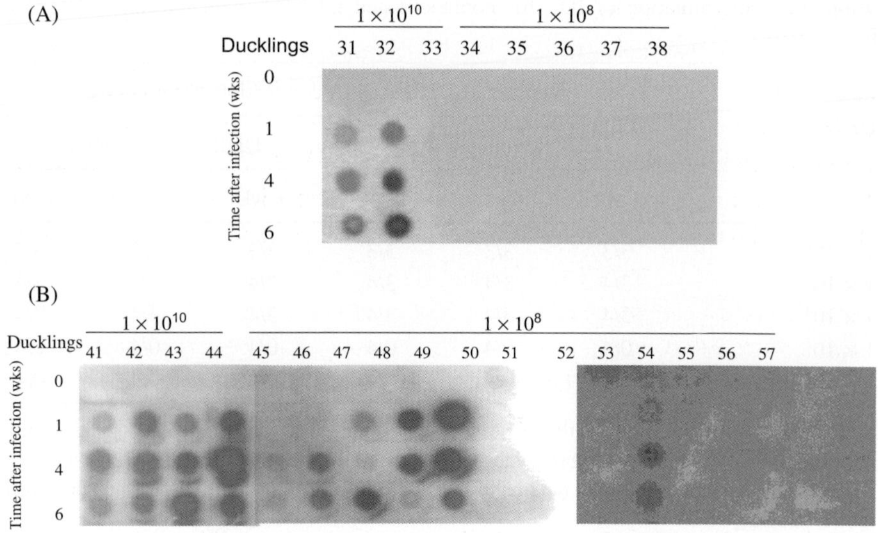


**Fig. 2.** Detection of DHBV DNA in the sera from ducklings inoculated with virus-inactivated DHBV in human plasma or human RBCs. One-day-old ducklings were intravenously infected with DHBV in different copies added to human plasma (DHBV–plasma) (A) or human red blood cells (DHBV–RBCs) (B). Sera were collected at intervals, and detected by dot hybridization with a DHBV DNA probe labeled with $^{32}P$. The sera from ducklings infected with $1 \times 10^3$ and $1 \times 10^2$ copies of DHBV were all negative. Therefore, we do not show the data here.

inoculated with $1 \times 10^8$ copies showed DHBV infection. $ID_{50}$ of MB-treated DHBV–human RBCs was $1 \times 10^{8.23}$ copies. Compared with $ID_{50}$ of untreated DHBV–human RBCs, MB treatment decreased the virus titer in human RBCs by about five logs.

To confirm the results from spot hybridization, total DHBV DNA was detected in the liver of the ducklings by Southern blot hybridization at the end of six weeks, and the results on duck livers detected were consistent with those in the sera (data not shown).

## 3. Discussion

In the process of blood donor screening, due to the limitations on technology or the lack of highly sensitive methods for virus detection, there is still no effective method to test for viral infection during the window period of infection, when the virus load is low and the immune response is still undetectable.[6,7] *In vitro* viral inactivations of blood components are, therefore, an effective measure to prevent posttransfusion infections of bloodborne viruses such as HBV, HCV and HIV. A variety of approaches to inactivation of viruses in the blood components in the clinic have been developed, such as heating at 60°C for 10 h, treatment of plasma fractions with the solvent–detergent (S/D; 1.0% detergent and 0.3% tri-*n*-butyl phosphate solvent) process, using photoactive compounds and visible light to inactivate for robust removal of viral activity, using β-propiolactone and with long-wave ultraviolet light, and the recently used MB photodynamic method. In the 1990s, it was proven that viruses could be inactivated by MB treatment with irradiation of visible light.[8] Since then its application for inactivation of viruses has drawn intensive attention, and MB treatment has been proven to efficiently inactivate both RNA and DNA envelope viruses.[9] MB, as a reducing agent, was mainly employed in thyroid gland imaging, relief of toxicosis from cyanide, and therapy for methemoglobinemia induced by nitrite or nitrobenzene. It is a cationic phenothiazine dye that can bind to viral lipid membrane and viral nucleic acids with higher affinity. After absorbing light energy, MB is excited and transfers the absorbed energy onto molecular oxygen-producing singlet oxygen, resulting in disruption of viral lipid membrane and destruction of nucleic acid, which can inactivate most envelope viruses and a number of nonenvelope viruses without affecting the quality of blood components.[10] In the 1990s, researchers from Europe, the USA and the Shanghai Blood Center began to develop the MB photodynamic method for treating single bag plasma to inactivate viruses, and it was proven to efficiently inactivate viruses in the plasma and to have no significant side effects on the virtual components of the plasma. The Shanghai Blood Center has developed the MB photodynamic method under the best conditions to inactivate the virus and it has shown high efficacy, as validated by the enveloped RNA viruses VSV and Sindibis virus. The method has been in use in clinics and its efficacy has been confirmed in a number of studies.

The previous assay for evaluating the efficacy of viral inactivation relied mainly on an *in vitro* infectivity assay with VSV, Sindbis virus or

HIV in cell culture. In the present study, we developed a method with DHBV, an enveloped DNA virus, and infected an animal model to evaluate DNA virus inactivation treated by the MB photodynamic method. The duck is a natural host of DHBV, and a duckling can be infected by inoculation of the virus in the vein. DHBV is a bloodborne virus belonging to the family Hepadnaviridae, which includes HBV.[11] The DHBV-infected duckling is an animal model for studies of HBV infection and viral replication. Therefore, it can be used to test the efficacy of viral inactivation.[12,13] No effect of human plasma and RBC inoculation alone on duckling growth and no ill-effect on ducklings were observed during the experiment period. Thus, this model is appropriate for evaluation of the effect of the MB photodynamic method on DNA virus inactivation.

As shown in this study, after MB photodynamic treatment, $ID_{50}$ of DHBV–human plasma decreased by six logs, while $ID_{50}$ of DHBV–human RBCs decreased by five logs. This indicated that MB photodynamic treatment could inactivate $1 \times 10^8$ copies in 200 μl of human plasma, while only 46% of $1 \times 10^8$ copies had been inactivated in 200 μl of human RBCs. MB photodynamic treatment could reach 100% of viral inactivation when the virus titer was low. In blood donors, the ones that fail to be identified are usually at low viral loads. Thus, we suggest that before transfusion, if the blood components are further treated with viral inactivation, it would be safer for the recipients.

## References


1. Barbara JAJ *et al.*, *Transfus Sci*, 1998, **19**: 3–7.
2. Guertler LG *et al.*, *Thromb Res*, 2002, **107**: S39–S45.
3. Qian KC *et al.*, *J Clin Transfer Lab Med*, 2001, **3**: 2–5.
4. Huang YW *et al.*, *Chin J Disinfect*, 2000, **17**: 1–4.
5. Santus R *et al.*, *Clin Hemorheol Microcirc*, 1998, **18**: 299–308.
6. The Public Health Service, *MMWR Morb Mortal Wkly Rep*, 1991, **40**(RR-4): 1–17.
7. Dodd RY, *N Engl J Med*, 1992, **327**: 419–421.
8. Horowitz B *et al.*, *Ann Med*, 2000, **32**: 475–484.

9. Wallis C *et al.*, *Photochem Photobiol*, 1966, **4**: 159–170.
10. Mohr H *et al.*, *Dev Biol Stand*, 1993, **81**: 177–183.
11. Summers J *et al.*, *Cell*, 1982, **29**: 403–415.
12. Eble BE *et al.*, *Photochem Photobiol*, 1998, **67**: 700–713.
13. Jilbert AR *et al.*, *Virology*, 1998, **244**: 273–282.

# 2.4

# Sequence Analysis of the SH Gene and Its Flanking Region of a Mumps Virus Strain (SP Strain)

Shao-Hui Ma*, Rong-Song Zhang†, Long-Ding Liu*, Jing-Jing Wang*, Li-Chun Wang*, Wen-Jun Yang‡, Yan Liang*, Zi-Jiang Yang* and Qi-Han Li*,§

**Institute of Medical Biology, Chinese Academy of Medical Sciences, Peking Union Medical College, Kunming 650118, China*
*†Honghe Prefecture Center for Disease Control and Prevention, Mengzi 661100, China*
*‡Shiping County Center for Disease Control and Prevention, Shiping 662200, China*

**Objective.** To study the genetic characteristics of the SH gene and its flanking region of the mumps virus strain SP by comparing sequences between this strain and the genes of other wild mumps viruses.

**Methods.** A saliva sample from a child suffering from mumps was collected in Shiping county, Yunnan province, in 2005, and inoculated in Vero cells. The cytopathic effect (CPE) was observed in the inoculated cells, and the virus in supernatant was harvested and identified by HA. Viral RNA was inoculated and reverse-transcribed into cDNA for PCR

§Corresponding author. Email: qihanli@21cn.com

amplification. This cDNA fragment (0.4 kb) was sequenced and analyzed by Vector NTI6.

**Results.** The SP strain showed a close genetic relationship (93%–96% homologous) with some strains isolated in other parts of China. Analysis of the amino acid sequences indicated that the SH sequence in the SP strain is 93.3%–98.25% homogenous with the Wlz2 strain.

**Conclusion.** The SP strain is a member of mumps virus genotype F.

*Keywords*: mumps virus SP strain; small hydrophobic gene; sequence analysis.

## 1. Introduction

The mumps virus belongs to the family *Paramyxoviridae* and to the genus *Rubulavirus*. The single-stranded mumps virus genomic RNA contains seven genes, in the following order on the genome map: the nucleocapsid (N), phospho (P), membrane (M), fusion (F), small hydrophobic (SH), haemagglutinin ± neuraminidase (HN) and large (L) protein genes. The SH gene of the mumps virus encodes a protein of 57 amino acids. A reduced or absent expression of the SH gene, with a concomitant reduction of the SH protein, has been described for certain mumps virus strains. The lack of expression of the SH protein does not affect virus replication *in vitro*, but may modify virus pathogenesis *in vivo*. The SH gene is the most variable part of the mumps virus genome. Since the mumps virus is monotypic, a vaccine from any strain should provide lifelong protection against subsequent infection. However, mumps outbreaks have not been completely eliminated even in vaccinated populations and, in some cases, mumps-virus-neutralizing antibodies do not protect against reinfection with a heterologous mumps virus genotype. Therefore, it is of interest to develop a novel mumps vaccine with improved immunity, efficacy and safety. A mumps virus clinical isolate, the SP strain, was obtained from a throat swab of a 13-year-old patient with parotitis in Shiping, China. To study the genetic characteristics of the SP strain, the SH gene and its flanking region of the mumps virus strain SP was sequenced and compared to the genes of other wild mumps viruses.

# 2. Results

## 2.1. *Amplification of the SH gene and its flanking region of the mumps virus*

The SH gene and its flanking region within the mumps virus were successfully amplified by RT-PCR with the designed primers. The fragments for the SH gene and its flanking region (413 bp) were obtained. The results from a PCR reaction are shown in Fig. 1.

## 2.2. *Phylogenetic analysis*

Phylogenetic trees, based on the entire SH gene (315 nt in length), were constructed using the neighbor-joining method. They are shown in Fig. 2. The SP strain formed a monophyletic group with strains UK99–102 × 13, UK99–190, Wlz1, Wlz2, Wlz3, Wsh1 and Wsh2. The SP strain was genetically distinct from other mumps virus strains isolated in other areas. However, it had the same ancestry and was closely related to Wlz1, Wlz2 and Wlz3, indicating that these strains may belong to the same genotype.

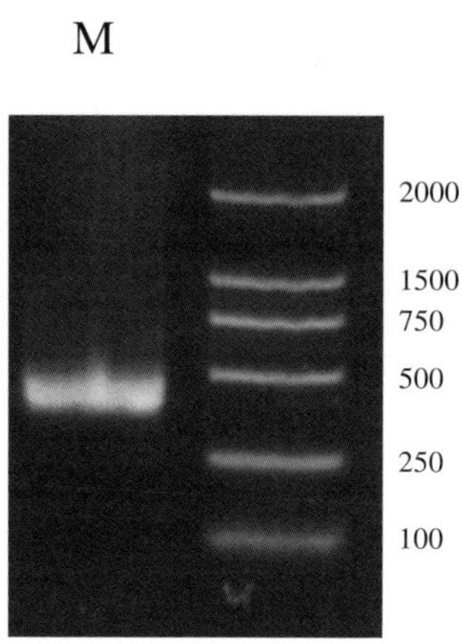


**Fig. 1.** The RT-nested PCR products of the SH gene and flanking regions of the mumps virus SP strains. M: DNA marker DL2000.

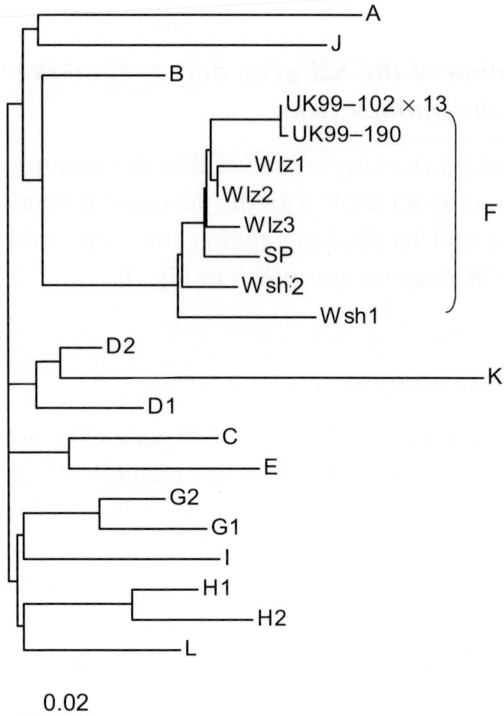


**Fig. 2.** Phylogenetic tree of the mumps virus.

Wu *et al.* proposed that 8% sequence diversity be used as a cutoff point for the delineation of new genotypes.[1] Compared with UK99–102 × 13, UK99–190, Wlz1, Wlz2, Wlz3, Wsh1 and Wsh2, the SP strain showed 93%–96% nucleotide similarity (Fig. 3). Because the UK99–102 × 13, UK99–190, Wlz1, Wlz2, Wlz3, Wsh1 and Wsh2 strains belong to the F genotype, it can be concluded that the SP strain also belongs to it.

## 2.3. *Analysis of the amino acid sequence of the SH gene*

The SP strain, in comparison with UK99–102 × 13, UK99–190, Wlz1, Wlz2, Wlz3, Wsh1 and Wsh2, showed a divergence ranging from 1.75% to 12.28% at the amino acid level. However, a relatively high divergence was observed in the UK99–102 × 13, UK99–190, Wsh1 and Wsh2 strains.

| | 1 50 |
|---|---|
| SP | mpaihppfyltflllillhliitlyvwimltithktavqhaalyqrslfr |
| UK99–190 | -------l-------v------------t---i---------------v- |
| Wlz1 | -s-----l---------------------------------g-------- |
| Wlz2 | -------l------------------------------------------ |
| Wlz3 | -------l--------------------------------vg-------- |
| Wsh1 | -s-----l---------------------------------v------rs |
| Wsh2 | -s-----l---------------s---------y---g------------ |
| UK99–102 × 13 | -------l-------v----------------i---------------v- |

| | 51 |
|---|---|
| SP | wsfdhsl |
| UK99–190 | ------- |
| Wlz1 | ------- |
| Wlz2 | ------- |
| Wlz3 | ------- |
| Wsh1 | -gl---- |
| Wsh2 | -g----- |
| UK99–102 × 13 | ------- |

**Fig. 3.** Comparison of deduced amino acid sequences of the SH gene in F genotypes.

A relatively low divergence was observed between the SP strain and Wlz2. These data indicate that the homology of the SP strain and Wlz2 is very high (Table 1).

## 3. Discussion

In the present study, a rapid and easy test for genotyping was developed using RT-PCR with genotype-specific primers. The method may be useful

**Table 1.** Comparison of cDNA sequences of the SH gene and flanking regions of F genotype strains.

| Position of cDNA sequences of the SH gene and flanking regions / Strain | Wlz3 | Wsh1 | Wsh2 | Sp | Wlz1 | Uk99–102 × 13 | UK99–190 | Wlz2 |
|---|---|---|---|---|---|---|---|---|
| 8(T) | – | G | – | – | – | – | – | – |
| 29(C) | – | – | – | – | T | – | – | – |
| 49(A) | – | – | – | – | C | – | – | – |
| 71(A) | – | – | – | – | – | G | G | – |
| 79(T) | – | C | – | – | – | – | – | – |
| 96(T) | – | C | – | – | – | – | – | – |
| 112(T) | – | – | – | – | – | – | C | – |
| 116(A) | – | G | – | – | – | – | – | – |
| 124(T) | C | C | C | C | C | – | – | C |
| 147(C) | T | T | – | – | – | – | – | – |
| 150(C) | G | – | – | – | G | – | – | – |
| 152(C) | – | – | – | T | – | – | – | – |
| 168(T) | – | G | – | – | – | – | – | – |
| 171(C) | – | C | – | – | – | – | – | – |
| 175(T) | – | C | – | – | – | – | – | – |
| 179(A) | – | G | G | – | – | – | – | – |
| 181(T) | – | – | – | – | – | C | C | – |
| 182(T) | – | C | – | – | – | – | – | – |
| 202(A) | – | C | – | C | – | C | C | – |
| 211(A) | – | G | – | – | – | – | – | – |
| 233(C) | – | – | – | – | – | T | T | – |
| 250(C) | T | – | – | – | – | – | – | – |

for genotype screening of a large number of specimens. However, to obtain more detailed information, it was necessary to perform a nucleotide sequence analysis of the gene encoding for the SH protein.[2,3]

Orvell *et al.* tested ten samples of human serum from children who had been vaccinated with the JL virus strain. The neutralizing antibody titers varied from 4 to 64 in tests with the JL strain. Neutralization titers were two- to eight-fold lower in tests with the SBL-1 strain. In tests with the Kilham and V6 virus strains, lower neutralization titers were found for the JL strain, but the difference was not as pronounced as with the SBL-1 strain.[4] Different criteria for reinfection were met, including the demonstration of IgG antibodies by ELISA in a preinfection serum sample. The preinfection serum sample from the patient was able to efficiently neutralize the infectivity of a heterologous genotype D strain, but was unable to neutralize the homologous genotype A virus. The findings in the present study may offer an explanation for a mechanism behind previously observed vaccine failures and the occurrence of reinfection with heterologous mumps virus strains.[5]

Although the current vaccine is effective, the mumps virus has at least 11 genotypes. Therefore, the development of a better mumps vaccine with improved effectiveness and safety, in light of the adverse events related to the current vaccine, is clearly needed.

The mumps virus, isolated from the Lanzhou, Shanghai and Tangshan, belongs to genotype F and shows an over 11% sequence divergence from all other genotypes.[1,6] We sequenced the SH gene and its flanking region of the mumps virus strain SP and compared it to the genes of other wild mumps viruses. The results showed that the SP strain shared a relatively high homology with the UK99–102 × 13, UK99–190, Wlz1, Wlz2, Wlz3, Wsh1 and Wsh2 strains. A relatively low divergence was observed between the SP strain and Wlz2. On the basis of the SH gene sequence analysis, it has been shown that those MuV isolates could belong to genotype F. The surveillance of the mumps virus in China should be improved in order to better understand the epidemiology of the virus in the southwestern region.

The results show that the mumps virus isolated from China mainly belongs to genotype F. At present, the live attenuated mumps vaccines (S79 and ME) in China belong to genotype A with Jeryl Lynn.[7,8] This study concludes that it is advantageous to obtain a new mumps vaccine candidate using genotype F.

## References

1. Wu L *et al.*, *Vaccine*, 1998, **16**(2–3): 281–285.
2. Jin L *et al.*, *J Infect Dis*, 1999, **180**(3): 829–833.
3. Kim SH *et al.*, *Microbiol Immunol*, 2000, **44**(3): 173–177.
4. Orvell C *et al.*, *J Gen Virol*, 2002, **83**(Pt 10): 2489–2496.
5. Nojd J *et al.*, *Vaccine*, 2001, **19**(13–14): 1727–1731.
6. Sun S *et al.*, *Chin J Coal Industry Med*, 2005, **8**(10): 1151–1153.
7. Zhang Y *et al.*, *Vaccine* (Science Press, Beijing) 2004, pp. 1248–1249.
8. Xu H *et al.*, *Chin J Microbiol Immunol*, 1999, **19**: 215–218.

# 2.5

# HIV-Related Knowledge, Attitudes and Behaviors Among Policewomen and Female Prisoners from a Behavior Correction Center

Wei Jiao*, Ying-Jie Zheng*, Xue-Cai Wang†, Ai-Qin Xu†, Jian-Ping Gu†, Jian-Fu Zhu†, Yi-Han Lu*, Fa-Di Wang† and Qing-Wu Jiang*,‡

**Key Laboratory of Public Health Safety, Ministry of Education, Department of Epidemiology, School of Public Health, Fudan University, Shanghai 200032, China*

*†Deqing County Center for Disease Control and Prevention, Deqing 313200, China*

**Objective.** To study the status of HIV-related knowledge, attitudes and behavior among female guards and female prisoners who have been arrested for prostitution and/or illegal drug use and to provide a scientific basis for prevention and control of HIV/AIDS among this group of people.

**Methods.** The female guards were divided based on their roles in the prison (i.e. administrators or ward workers), while the female prisoners were analyzed based on the reason for their incarceration (i.e. prostitution or drug use). Focused group interviews ($n$ = 8/group) were conducted

‡Corresponding author. Email: jiangqw@shmu.edu.cn

and AIDS-related education with basic knowledge of AIDS/HIV was provided to the female guards and female prisoners, separately.

**Results.** Our study found that, in general, the guardswomen held a discriminating attitude toward the female prisoners. The female drug users usually shared needles and had a distrust of the government's program of syringe exchange. The condom use rate for the female sex workers was high during sexual intercourse with their customers, but low during sexual intercourse with personal partners. In contrast to those that lived in large metropolitan areas, women prisoners from small towns and cities did not have adequate health education or social rectification systems.

**Conclusion.** The discrimination against the female prisoners shown by the guardswomen should be eliminated through health education. Health education in the community should be strengthened, especially in small towns and cities. Simultaneously, more attention and concern should be given to HIV-related high-risk behaviors.

*Keywords*: acquired immunodeficiency syndrome; female sex workers; qualitative study; behavior.

## 1. Introduction

Currently, the epidemiology of acquired immunodeficiency syndrome (AIDS) in China is fast-growing. New cases of infection have arisen mainly via intravenous drug use and sexual transmission. HIV-1 infection rates for drug users and prostitutes increased to 6.5% and 1%, respectively, in 2004, up from 2% and 0.02%, respectively, in 1996,[1,2] indicating that AIDS is increasing in this group of people with high-risk behaviors. This increase in HIV-1 infections among this vulnerable group has the potential to spread to the segment of the population with low- or no-risk behaviors, if effective control measures are not taken.

Notably, female sex workers and drug users play a major role in the AIDS epidemic.[3] Education regarding AIDS prevention at the many correction centers in China should be extensively and persistently scheduled in order to increase the knowledge about AIDS prevention among prisoners, in order to change their high-risk behaviors, to increase self-protection, and to prevent the spread of AIDS among prisoners. This is an important task for AIDS prevention and for keeping the infection rate under control at the present time and in the future.[4] Furthermore, education at the correction centers will directly affect the attitude of guardswomen toward prisoners, which will increase the behavior change effects among prisoners.[5] This research was conducted to investigate the attitudes, behaviors and knowledge regarding HIV/AIDS among guardswomen and prisoners from a female behavior correction center through group interviews, with the goal of providing evidence for the necessity of AIDS prevention/control among a group of incarcerated prostitutes and drug users.

## 2. Results

### 2.1. *Basic information on interviewees*

The guardswomen in this sample were aged 24–34 and came from different police squads; the administrative group ranged in age from 29 to 48, and included vice supervisors, secretaries from medical departments, political directors from sexually transmitted diseases departments, and directors from different administrative departments. The prostitutes ranged in age from 19 to 38 and were from Hubei, Jiangxi, Henan and Jilin provinces. They provided commercial sexual services mainly in hotels, KTV lounges and beauty salons. Many of them had gone to primary or secondary school. The drug users ranged in age from 22 to 32 and were from Zhejiang, Hunan, Anhui and Sichuan provinces. They had gone to secondary or high school, or secondary technical school.

### 2.2. *Knowledge about AIDS among the policewomen and female prisoners*

#### 2.2.1. *Resources for gaining knowledge*

AIDS-related knowledge among the policewomen and female prisoners was gained mainly via the media, including TV, newspapers and other public channels and booklets. Prisoners commented that information on

AIDS had been frequently distributed and highly publicized in the society. Particularly, information on AIDS was available to sex workers in their places of employment in big cities. Knowledge of AIDS and sexually transmitted diseases was disseminated throughout the probationary period of the prisoners, upon their entering the center, during their routine education and upon their release from the center. Effects of education were successfully achieved through various activities within the prison, including booklet distribution, health-related knowledge and speech contests, information in blackboard posters, video broadcasts and prisoner-presented lectures. While systematic lectures given by guardswomen were informative for the prisoners, the content of the lectures was based too much on theory and insufficiently on real-world experience. Prisoner-presented speeches (peer group education) were based more on real-world experience and made a greater impression on fellow prisoners. The consensus among prisoners was that combining these two types of education would be more helpful.

### 2.2.2. *Attitude toward AIDS*

Knowledge of AIDS among the guardswomen, particularly those in administrative roles, had more scientific details, and all of them were aware that the drug users and prostitutes at the correction center were at risk of being infected with HIV. Though some of the policewomen indicated that the prisoners were dealt with fairly and without discrimination regarding effective education, there was discrimination, to some extent, among the guardswomen, who believed that the behavior of the prisoners would not be changed by education and, therefore, did not see any reason to provide them with AIDS-related education.

Although the prisoners expressed fear about contracting AIDS, they believed that infection was not a risk since they were not in contact with AIDS patients. Only one prisoner, from Anhui, said that "although I was able to take measures to keep myself safe from contracting AIDS, it does not guarantee that my boyfriend or husband is free from AIDS. AIDS has spread worldwide, so it is very close to us." All the prostitutes said that they would not provide sexual services once infected with HIV, but they also said that there was high social pressure on them and that their psychological tolerance was limited.

### 2.2.3. *High-risk behaviors leading to AIDS*

Strong self-protection measures were taken by all the guardswomen, particularly those who were working closely with the prisoners on a daily basis. These measures included wearing gloves for all contact with prisoners, using disinfectants, and not having direct contact with prisoners' blood and/or secretions during daily operations and education.

Most of the prostitutes who were interviewed said that condoms were used during sexual services, and that if the use of a condom was refused by a client, he would not be served. However, a number of them reported that some of their friends still provided sexual services for money without the use of a condom. Despite practicing safe sex during their prostitution activities, most women reported that they did not use condoms during intercourse with their husbands or boyfriends. The prostitutes were concerned about sexually transmitted diseases and they often underwent medical examinations at one- or three-month intervals, indicating that, unlike in the past, they were no longer ashamed of receiving medical attention when they were ill.

Various types of clinics were chosen by the intravenous drug users in the sample for buying syringes. The reuse of syringes after they had been treated with alcohol was common among most of the drug users. A few prisoners incorrectly believed that water at over 30°C could be used for sterilization, and none of the prisoners knew of methods that were suitable for disinfection. Syringe-sharing with a boyfriend or stable drug user partner was admitted in over half of the prisoners. The present investigation showed that the drug users in this group did not partake in providing sexual favors and that the source of drugs was the prisoners' families or boyfriends.

## 2.3. ***Need for health knowledge and intervention***

Due to a lack of medical knowledge about syringe-cleaning procedures among the prison staff, the prisoners were not informed of these techniques and were only told not to share their needles.

The prostitutes in this sample were very interested in educating themselves about AIDS and some of them reported buying books for self-education. Sexually transmitted diseases were detected at a higher rate at the correction center, but because of a shortage of laboratory diagnostic workers, medical examination for prisoners was done only upon their entering the center, with no re-examination for those with a negative test result. Infected and negative prisoners were managed separately, and were not allowed to share a bed.

Methadone alternative therapy can be chosen by prisoners in big cities if they do not have a history of compulsive drug-taking. The prisoners were willing to accept methadone therapy without heroin, but a few of the guardswomen disagreed about the effect of methadone because of the difficulty of stopping it. There was common skepticism about free syringe exchange programs provided by the government for prisoners, who indicated that these programs appeared to promote injection drug abuse. Only a few of the prisoners believed that it was an approach to preventing the spread of AIDS. It was also a common concern that they would be arrested when exchanging syringes, and only if syringes were unavailable would they consider using syringe exchange and doing so with the help of friends who were not drug users. Urine tests and help-education systems were established in some of the provinces for controlling relapse. It was thought by prisoners from other areas that if better urine-testing procedures and help-education systems were available in local areas, the rate of relapse among drug users might decrease.

## 2.4. *Effect of social concern*

In the present investigation, all drug users initiated their drug use when they were younger than 20 years old. It was suggested by an administrative prison worker that the government should make a greater effort to educate middle school students on AIDS-related issues, because most of the prisoners had been educated in secondary schools when they started using drugs. If information on AIDS had been popularized in the middle schools, it is possible that many of the current drug users would not have started using drugs, resulting in a decrease of drug users and prostitutes. Education in the secondary or high schools would provide students with knowledge about AIDS and drug abuse, which would allow them to protect themselves. As part of their rehabilitation, the female prisoners in the current study reach out to the community and provide information on their experiences with drugs and/or AIDS. These activities include speeches given by prisoners in schools, public distribution of AIDS information, press releases, video broadcasting, and consulting services by young guardswomen during Youth Day (on the fourth of May). The administrative workers in the prison indicated that AIDS education was extensive and persistent at the correction center, but was rarely available to the public. Advice on using condoms was prohibited at the center, because this would mean acquiescence and encourage nonrectification. Many commented that the owners of places providing commercial sexual services should take some of the social responsibility as no warning about AIDS is given by these owners, and innocent women are encouraged to provide sexual services. One prostitute stated that she was reprimanded for not providing sexual services to a customer who refused to use a condom.

## 3. Discussion

Due to the confidential nature of the interviews and because guardswomen were not allowed to be present during an interview with a prisoner, we believe that the opinions expressed here reflect the true opinions of prostitutes and drug users from various provinces of China on AIDS-related issues.

When compared to previous research,[6] the present results reveal that the guardswomen had a better understanding of AIDS-related issues, evidenced by their awareness of the importance of self-protection and their knowledge of how to conduct such measures if necessary. The extent of knowledge displayed by the guardswomen had a direct effect on that of the prisoners. Several issues were revealed in this investigation, indicating that AIDS-related education should be improved among guardswomen, including correct syringe-cleaning procedures, a nondiscriminating attitude to prisoners, etc. Meanwhile, AIDS-related knowledge among prisoners was gained to some extent from society. We found that some progress in public education had been made, with more emphasis on personal health care and the importance of the use of condoms during sexual intercourse.[7] We also found that basic health care departments in small hospitals located in small cities and towns played an important role in preventing AIDS in certain populations. One trend of particular concern was that although prostitutes reported the use of condoms during sexual intercourse with clients, they also said that they did not use condoms during sexual intercourse with their fixed sexual partners.[8,9] Since syringe-sharing was common among drug users, AIDS-related knowledge should be promoted in this group.[10] The present research was performed in a region of China with a low incidence of AIDS,[11] and in comparison with larger cities, public education and behavioral interventions were less common.[12] Therefore, we suggest that AIDS-related education should be enhanced and that methadone therapy should be provided to individuals in smaller communities,[13] and that a complete health education system should be established,[14] so as to provide the residents of these smaller towns and cities with an education-rich environment.[15]

Discrimination against prisoners was common among guardswomen who were in frequent contact with prisoners.[5] Besides, disbeliefs and suspicions about the government and society are still common among drug users and prostitutes, and they are shown by the negative attitudes toward the free syringe exchange programs provided by the government.[16] Therefore, aside from educating about AIDS and drug abuse, information regarding the policies of the government, and social attitudes toward other issues, should be included at the education center.

AIDS prevention was often ignored by owners of KTV lounges, beauty parlors, and other places where sexual favors are offered for a fee. Constant on-site inspection and education for these owners should be strengthened to change their attitudes[17] and to gain their cooperation. AIDS education should be conducted for students of a young age, including those in middle school, which may be crucial for prevention of AIDS.[18] Investigations similar to the present one, among different populations on different health problems in China, will yield firsthand information for policymakers, and in the end will benefit all members of society.

## References


1. Gao Y, *Chin Health Educ*, 2003, **19**(9): 703–704.
2. The Health Department of the People's Republic of China, *Progress of AIDS Prevention and Treatment in China in 2005* (Beijing, 2005).
3. Belenko S *et al.*, *AIDS Educ Prev*, 2004, **16**(4): 367–385.
4. Xu SL *et al.*, *J Med Pharmacol Symp*, 2005, **26**(19): 9–10.
5. Yang QY *et al.*, *Pract Prev Med*, 2005, **12**(5): 1114.
6. Wang P *et al.*, *Soft Sci Health*, 2001, **15**(6): 12.
7. Du JW *et al.*, *Chin J Tropic Med*, 2005, **5**(8); 1634.
8. Lu J *et al.*, *Bull Chin Edem*, 2005, **20**(4): 23.
9. Zhu GY *et al.*, *Chin AIDS Sex Transm Dis*, 2003, **9**(2): 95.
10. Xu YS *et al.*, *Chin J Sex Transm Dis AIDS Prev*, 2001, **7**(4): 230.
11. Yang JZ *et al.*, *Zhejiang J Prev Med*, 2005, **17**(11): 13.
12. Zhang XW *et al.*, *Chin Health Educ*, 2001, **17**(7): 397.
13. Liu ZQ *et al.*, *Dis Surv*, 2005, **20**(2): 68.
14. Zeng LM, *Chin J Drug Abuse Prev*, 2001, **33**(4): 15.
15. Wang W *et al.*, *Acta Zhejiang Ind Comm Univ*, 2005, **74**(5): 51.
16. Luo J *et al.*, *Chin Health Educ*, 2006, **22**(2): 115.
17. Zhou YJ *et al.*, *Chin AIDS Sex Transm Dis*, 2003, **9**(3): 170.
18. Guo JW *et al.*, *Soft Sci Health*, 2002, **16**(5): 40.

SECTION 3

# Laboratory Medicine

# Introduction and Comments: Advances and Trends in Diagnostic Microbiology in China

Yi-Wei Tang

*Departments of Pathology and Medicine,*
*Vanderbilt University Medical Center,*
*Nashville, TN 37232, USA*

The famous Canadian physician William Osler pointed out in *The Principles and Practice of Medicine: Designed for the Use of Practitioners and Students of Medicine*, his medical textbook published in 1892, that "the three most important elements in practicing medicine are diagnosis, diagnosis and diagnosis." In the field of diagnostic microbiology, laboratorians or clinical microbiologists determine whether pathogenic microorganisms are present in clinical specimens collected from patients with suspected infections. Overall, a pathogenic microorganism from a test sample can be detected and identified in any of four possible ways: (1) cultivation of microorganisms using artificial media or living hosts; (2) direct microscopic examination; (3) measurement of microorganism-specific immune responses; and (4) detection of microorganism-specific macromolecules, especially nucleic acids. During the past two decades, technical advances in diagnostic microbiology have made constant and enormous progress in various areas, including bacteriology, mycology, mycobacteriology, parasitology and virology. The diagnostic capabilities of modern clinical microbiology laboratories have improved rapidly and have expanded greatly due to a technological revolution in molecular aspects of microbiology and immunology. In particular, rapid techniques for nucleic acid amplification and characterization combined with automation and user-friendly software have significantly broadened the diagnostic arsenal for the clinical microbiologist. These techniques have appeared in China as well, except that all of them did so in a much shorter period of time within the past ten years.

Several articles, which were previously published in the *Journal of Microbes and Infections* and are collected in this book, provide exemplary recent translational studies in the field of diagnostic microbiology in China. In addition, they exactly reflect the technical changes. After 1979, conventional diagnostic technology evolved rapidly. Automatic, continuous monitoring blood culture systems have been implemented in large hospitals in China. Immunoassays came into wider use in both antigen and antibody detection in the diagnosis and monitoring of viral hepatitis. Application of monoclonal antibodies has significantly increased the sensitivity and specificity of antigen detection. Zhou *et al.* (Chapter 3.1) from Shanghai Public Health Clinical Center affiliated to Fudan University developed two immunoassays to detect SARS-associated coronavirus-specific antibodies. Wang's group (Chapter 3.7) in the Department of Microbiology of Anhui Medical University developed a modified enzyme-linked immunosorbent assay to measure the human cytomegalovirus (HCMV)–specific pp65 IgM and IgG avidity index, and to use the index to differentiate between HCMV primary infection and reactivation. A microhemagglutination inhibition test was used by Qin *et al.* (Chapter 3.6), from the Guangzhou Center for Disease Control and Prevention, to study the epidemiological features of human infection with avian influenza. Gao *et al.* (Chapter 3.8), from the same institute, analyzed the patterns of indeterminate HIV Western blot results. Fifty percent of the indeterminate results were from relatively healthy people, including blood donors, pregnant women and those who voluntarily asked for counseling and testing. Most of these indeterminate results turned out to be false positive cases, due to the p24 band.

PCR-led *in vitro* nucleic acid amplification technologies have significantly broadened the microbiologist's diagnostic arsenal. There was no exception in China in the development of new molecular methods for detection and characterization of pathogenic organisms. In the same article mentioned above, Zhou *et al.* also reported the use of a reverse-transcriptase PCR for early diagnosis of SARS-associated coronavirus infections. Wu *et al.* (Chapter 3.9), from Soochow University performed PCR on DNA extracted directly from acid-fast stain-positive sputum smears to detect rifampin resistance-related mutations in the *rpoB* gene. With a sensitivity of 91.7%, this assay may allow rapid organism detection and resistance determination since *Mycobacterium tuberculosis* grows slowly and requires several weeks to detect in clinical specimens using standard culture techniques. Meng *et al.* (Chapter 3.2), from Huashan Hospital, Fudan University were able to adapt advanced molecular techniques to further characterize an unusual *Brucella* species related to a sporadic brucellosis case in Shanghai. Zhou from Dalian Red Cross Center and Zhang from Dalian Medical University (Chapter 3.3) have adapted a PCR–capillary-electrophoresis method to detect and differentiate biovars 1 and 2 of *Ureaplasma urealyticus*. The assay can be a useful diagnostic tool since biovar 2 was demonstrated to be more associated with nongonococcal urethritis in males by this study.

Personalized medicine is emerging in the field of diagnostic microbiology and infectious diseases. If infections, especially chronic ones, can be viewed as "horizontally acquired" genetic diseases, it makes sense to view pathogen and host as an integrated system. Increasing evidence indicates that the outcome of many acute and chronic infectious diseases is influenced by the genetic background of the host. Host gene polymorphisms can modulate innate and/or acquired immunity that is associated with host susceptibility or resistance to certain infectious diseases by encoding altered gene products or by affecting transcriptional regulation. Two articles related to the field appear in the book. Zhang and Chen (Chapter 3.4) from the Children's Hospital of Fudan University studied expression levels of host leukocyte surface antigens CD64 during neonatal septicemia. The level of CD64 in the sepsis group was significantly higher than that in the other groups, indicating that the CD64 level can be used as an alternative, host biomarker for neonatal sepsis diagnosis. Yu's group (Chapter 3.5) at Xiangya Medical School of the Central South University used a PCR-based method to analyze the relationship between several MICA alleles and *Chlamydia trachomatis* infections. The data indicated that individuals with MICA*A5.1 allele may have a lower susceptibility to *C. trachomatis* infections. We look forward to more studies that will provide additional host biomarkers in the diagnosis and monitoring of infections.

Although significant advances have been achieved in China in clinical microbiology techniques, most of the technical developments were used for research projects and clinical trials and have not been implemented in routine clinical services. The old diagnostic microbiology model, which is either labor-intensive or requires days to months for test results to become available, remains the mainstay in most hospital clinical microbiology laboratories in China. Cell-culture-based viral organism detection and identification has been provided at only a very few diagnostic centers in large hospitals. Mass-spectrometry-based microbial identification and characterization has not been used in China for routine diagnostic services. Most molecular techniques are in the research-and-development stage, and limited molecular assays have been approved by the China State Food and Drug Administration and applied in routine diagnosis services. Beyond techniques, infrastructural and management construction has been carried out in an unbalanced manner. Several laboratories in prestigious university-affiliated hospitals have been successfully accredited by the College of American Pathologists, while the majority of clinical laboratories in small hospitals and clinics do not have written regulatory requirements. A quality management system needs to be in place for continuously analyzing, improving and re-examining resources, processes and services within a diagnostic microbiology service. A proficiency test program has been provided at the national level, but legislative actions related to failures are not implemented. Standardization of data management, methods, facilities, equipment, quality assurance, education and information systems at the national level has not been launched.

Diagnostic microbiology has been developing rapidly to detect and accurately identify implicated microorganisms in test specimens through a variety of techniques. Technological changes have made constant and enormous progress in various areas, including bacteriology, mycology, mycobacteriology, parasitology and virology, during the last two decades in the field. The implementation of nucleic-acid-amplification-based molecular techniques provides complementary, rapid and on-demand diagnosis services. These changes will continue, and inevitably lead to diagnostic microbiology being a modern discipline which can face any challenge worldwide, including China. We are expecting more translational researches to be carried out in this field in China, and looking forward to including more outstanding research findings in the next edition of this book.

3.1

# Detection of SARS-Associated Coronavirus Infection by Reverse Transcriptase-Polymerase Chain Reaction, Indirect Immunofluorescence Assay and Enzyme-Linked Immunosorbent Assay

Zhi-Tong Zhou*, Wei-Lun Jiang, Zheng-Hong Yuan, Fang Shen, Di Qu, Shi-Min Gu, Long-Ting Fu and Ren-Fang Zhang

*Shanghai Public Health Clinical Center affiliated to Fudan University, Shanghai 201508, China*

**Objective.** To study the use of molecular, biological and serological methods in diagnosis of SARS cases.

**Methods.** Samples from clinically diagnosed SARS patients, suspected patients and close contacts were studied by RT-PCR, IFA and ELISA.

**Results.** All three methods showed high specificity for detecting SARS CoV infection. Seven of the eight clinically diagnosed SARS cases were confirmed by specific laboratory results (among them, a case due to a positive IFA result was identified as clinically diagnosed SARS from a suspected patient). Anti-SARS CoV antibody was found positive in another suspected patient.

---

*Corresponding author. Email: zhouzt53@sina.com

**Conclusion.** The sensitivity of early diagnosis of SARS by RT-PCR, IFA and ELISA needs further improvement, but the combined use of these three methods could prevent misdiagnosis.

*Keywords*: severe acute respiratory syndrome; reverse transcriptase-polymerase chain reaction; indirect immunofluorescence assay; enzyme-linked immunosorbent assay.

## 1. Introduction

Severe acute respiratory syndrome (SARS) was characterized as an outbreak of an emerging infectious disease worldwide. It was first reported in the southern region of China in 2002. SARS was then identified as a respiratory viral illness caused by a new type of coronavirus, named SARS-associated coronavirus (SARS-CoV) in April 2003 by the World Health Organization (WHO). The methods of detecting SARS-CoV, including reverse transcriptase-polymerase chain reaction (RT-PCR), indirect immunofluorescence assay (IFA) and enzyme-linked immunosorbent assay (ELISA), have been established and played important roles in SARS diagnosis. But the limitation of these methods still means that they cannot meet the need for early diagnosis of SARS. Therefore, the WHO suggested that the epidemiology and clinical findings are the most important items for SARS diagnosis, and that laboratory diagnosis should be improved. From March to June of 2003, Shanghai Infectious Disease Hospital was in charge of the treatment of SARS patients and suspected SARS patients in Shanghai. A number of specimens from these patients and close contacts were collected and all specimens were assayed by RT-PCR, IFA and ELISA.

## 2. Results

### 2.1. *RT-PCR assay for SARS-CoV RNA*

A total of 102 specimens, including sera, throat swabs, throat washes and urine samples, were collected from SARS patients, or suspected SARS patients, or close contacts with SARS, and are shown in Table 1. The RT-PCR

**Table 1.** The collected specimens for detection of SARS-CoV by RT-PCR.

| Specimen from patients (number) | Specimens | | |
|---|---|---|---|
| | Sera | Throat washes | Urine |
| SARS patients (6) | 14 | 5 | 8 |
| Suspected SARS patients (32) | 37 | 19 | 1 |
| Close contacts (13) | 16 | 2 | – |

**Table 2.** The results of RT-PCR on 102 various samples from 51 subjects.

| Clinical group | Subject number/ positive number | SARS-CoV RNA[a] | | | |
|---|---|---|---|---|---|
| | | Sample type | Sample number | Positive number | Negative number |
| Confirmed cases | 6/5 | Serum | 14 | 3[b] | 11[c] |
| | | PWF[e] | 5 | 2[d] | 3 |
| | | Urine | 8 | | 8 |
| Suspected cases | 32/0 | Serum | 37 | | 37 |
| | | PWF | 19 | | 19 |
| | | Urine | 1 | | 1 |
| Close contacts | 13/0 | Serum | 16 | | 16 |
| | | Pharyngeal swab | 2 | | 2 |
| | | Urine | 0 | | 0 |

[a] Using kits produced by Kehua Bio-technical Co. Ltd., Shanghai. [b] Using kits produced by Artus Co., Germany. Two of these 3 serum samples were also positive (the remaining one was not tested). [c] Using kits produced by Artus Co., Germany. Of these 11 serum samples, 1 was positive and 1 was negative (the remaining 9 were not tested). [d] The 2 positive serum samples were detected using both Kehua's and Artus' kits. [e] PWF: pharynx washing fluid.

results on the 102 samples from 51 patients are shown in Table 2. Of the 6 clinically confirmed SARS patients, 5 were SARS-CoV RNA–positive. Only 3 of them were serum-positive; 1 was only throat-wash-positive, and 1 was both serum- and throat-wash-positive (these 4 positive serum samples were collected at 3, 6, 9 and 12 days after the onset of the disease, respectively; 2 throat-wash-positive samples were collected at 6 and 12 days after the onset). The SARS-CoV RNA of the other clinically confirmed SARS patient and the rest of the samples were all negative.

## 2.2. *IFA assay for SARS-CoV IgG antibody*

The results on 41 serum samples by IFA from 32 subjects are shown in Table 3. Among the 12 serum samples from 8 clinically confirmed SARS patients, 11 samples from 7 patients were SARS-CoV IgG antibody–positive. The other clinically confirmed SARS patient and the rest of the samples were all SARS-CoV IgG antibody–negative. The 12 SARS-CoV IgG antibody–positive samples from the 8 clinically confirmed patients are shown in Table 4.

**Table 3.** The results on 41 serum samples by IFA from 32 subjects.

| Clinical group | Subject number/ positive number | Anti-SARS-CoV IgG | | |
|---|---|---|---|---|
| | | Sample number | Positive number | Negative number |
| Confirmed cases | 8/7 | 12 | 11 | 1 |
| Suspected cases | 8/0 | 13 | | 13 |
| Close contacts | 15/0 | 15 | | 15 |
| Veterinary workers | 1/0 | 1 | | 1 |

**Table 4.** The dynamic results on serum anti-SARS-CoV IgG by IFA in eight confirmed SARS cases.

| Case | Days of onset and detection results | | | | | | | | | | |
|---|---|---|---|---|---|---|---|---|---|---|---|
| | 1–5 | 6–10 | 11–15 | 16–20 | 21–25 | 26–30 | 31–35 | 36–40 | 41–45 | 46–50 | 51–60 |
| 1 | | + | | | | + | | | | | + |
| 2 | + | | | + | | + | | | | | |
| 3 | | | | | | | | | | + | |
| 4 | | | | | | | | + | | | |
| 5 | | | | + | | | | | | | |
| 6 | | | | | | + | | | | | |
| 7 | +* | | | | | | | | | | |
| 8 | | | | – | | | | | | | |

*The serum sample of case 7 sent from Shanghai Pneumonic Disease Hospital; days of onset unknown.

**Table 5.** The results on 71 serum anti-SARS-CoV by ELISA from 35 subjects.

| Clinical group | Subject number/ positive number | Anti-SARS-CoV | | |
|---|---|---|---|---|
| | | Sample number | Positive number | Negative number |
| Confirmed cases | 8/7 | 41 | 30 | 11 |
| Suspected cases | 14/1* | 15 | 2 | 13 |
| Close contacts | 13/0 | 15 | | 15 |

*Two suspected cases' sera (collected on day 9 and day 16 after the onset) were reproducibly positive and their positive S/CO was 3.02 and 5.25, respectively.

## 2.3. *ELISA assay for SARS-CoV antibody*

SARS-CoV antibody of 71 serum samples from 35 subjects was detected by ELISA (Table 5), and the kinetic changes of anti-SARS-CoV antibody in serum samples from 8 clinically confirmed patients were detected (Table 6).

## 2.4. *Comparison of three different detection methods*

The results on 8 clinically confirmed SARS patients by three different methods are shown in Table 7. Among the 6 clinically confirmed patients tested by RT-PCR, all were positive except for patient No. 3 (serum and throat washes were both negative-assayed by two different kits). For the antibody test on 8 clinically confirmed patients, 7 were positive (except for patient No. 8).

Our results indicate that except for one, SARS-CoV antibody was detected for the 8 clinically confirmed SARS patients. Besides, SARS-CoV antibody was positive for one of the 14 suspected patients. We collected the serum of this suspected patient at different time points to recheck, and all samples were positive. The time point for the first detection of anti-SARS-CoV antibody was between 3 and 25 days after the disease onset.

# 3. Discussion

Our results indicate that the methods we used in this study, including RT-PCR, IFA and ELISA, had high specificity in testing SARS-CoV RNA

**Table 6.** The dynamic change of serum anti-SARS-CoV (total) by ELISA in eight confirmed SARS cases.

| Case | Days of onset and detection results (S/CO results) | | | | | | | | | | | | | |
|---|---|---|---|---|---|---|---|---|---|---|---|---|---|---|
| | 1–5 | 6–10 | 11–15 | 16–20 | 21–25 | 26–30 | 31–35 | 36–40 | 41–45 | 46–50 | 51–55 | 56–60 | 61–70 | 71– |
| 1 | <1/– | | 3.0/+ | | 4.3/+ | 2.1/+ | | | | 2.7/+ | 4.7/+ | | | |
| 2 | <1/– | <1/– | | <1/– | 1.4/+ | | 1/+ | | | | | | | |
| 3 | | 1.1/+ | | 5.5/+ | 2.6/+ | | 6.8/+ | 2.7/+ | | 2.5/+ | 4.3/+ | | 2.7/+ | 2.6/+ |
| 4 | | <1/– | 1.6/+ | | 1.9/+ | 4.0/+ | 5.5/+ | | 11.5/+ | | 18.0/+ | | | 7.5/+ |
| 5 | | <1/– | 20.3/+ | 18.6/+ | | | | | | | | | | |
| 6 | 2.1/+ | <1/– | 1.1/+ | 2.4/+ | | 2.3/+ | | | | | | | | |
| 7* | 7.6/+ | | | | | | | | | | | | | |
| 8 | <1/– | <1/– | <1/– | <1/– | | | | | | | | | | |

*The serum sample of case 7 sent from Shanghai Pneumonic Disease Hospital; days of onset unknown.

**Table 7.** Comparison of three tests on the various samples from eight confirmed SARS cases.

| Cases | RT-PCR | | | IFA (serum) | | ELISA (serum) | |
|---|---|---|---|---|---|---|---|
| | Days of onset | Sample type | Result | Days of onset | Result | Days of onset | Result |
| 1 | 10 | Serum | +[a,b] | 27 | + | 14 | + |
| 2 | 3 | Serum | −[a]/+[b] | 3 | + | 25 | + |
| 3 | 10 | Serum | −[a,b] | 47 | + | 9 | + |
| 3 | 10 | PWF[d] | −[a,b] | | | | |
| 4 | 9 | Serum | +[a,b] | 39 | + | 15 | + |
| 5 | 12 | PWF | +[a,b] | 20 | + | 14 | + |
| 6 | 6 | Serum | +[a]/−[b] | 29 | + | 3 | + |
| 6 | 6 | PWF | +[a,b] | | | | |
| 7[c] | | Serum | | ? | + | ? | + |
| 8 | | Serum | | 18 | − | 18 | − |

[a] Using kits produced by Kehua Bio-technical Co. Ltd., Shanghai. [b] Using kits produced by Artus Co., Germany. [c] The serum sample of case 7 sent from Shanghai Pneumonic Disease Hospital; days of onset unknown. [d] PWF: pharynx washing fluid.

and SARS-CoV antibody. Seven of 8 clinically confirmed patients and 1 suspected patient yielded positive results, while other, nonclinically confirmed subjects were all negative. According to the results assayed by the three different methods, 5 of the 6 clinically confirmed SARS patients were SARS-CoV RNA–positive by RT-PCR from their serum and/or throat washes at 3–12 days after the onset of the disease. Though SARS-CoV RNA was not detected in the serum or throat wash of patient No. 3, SARS-CoV antibody from his convalescent period serum was detected. Patient No. 7 had been excluded from suspected SARS infection in Hong Kong. While we did not have the RT-PCR results from his early samples, we detected anti-SARS-CoV antibody in his serum during the convalescent period, and subsequently he was confirmed as a SARS patient by Shanghai Municipal Center for Disease Control and Prevention (Shanghai CDC), and this positive SARS-CoV antibody result played a pivotal role in the SARS confirmation of this suspected patient.[1] We failed to detect anti-SARS-CoV antibody in the serum of patient No. 8 by day 18 after the onset of the disease.

We found a suspected SARS patient, whose serum samples collected at two different time points were all positive by the ELISA assay, although he had been excluded from SARS according to the clinical diagnosis. Whether this was false positive awaits further study.

Though many reports on the laboratory diagnosis of SARS-CoV were based on IFA and ELISA,[2–6] there are still controversies regarding the time point of antibody appearance, the order of specific IgG, IgM appearance, the implications of the antibody responses, etc. Lack of comparable results among reagents used in different laboratories makes it difficult to understand host immune responses to SARS-CoV infection.

Because SARS is an emerging infectious disease, the course of the disease has developed quickly, and it is highly contagious, with high mortality and high morbidity. Therefore, early isolation and early treatment is important for the control of this disease. Regrettably, till now, the sensitivity and specificity of laboratory diagnosis for SARS-CoV still cannot meet the clinical need. Although RT-PCR can directly detect the nucleic acid of the pathogen with high specificity, negative results of RT-PCR cannot exclude SARS, because the time of sample collection, the quality of samples, etc., can affect the results. The positive rates of anti-SARS-CoV antibody by IFA and ELISA are very low in the early period. According to the report of the Beijing group for national SARS prevention and treatment of emergency technology, sera from 267 SARS patients and 132 control sera were assayed by IFA and ELISA, and the results suggested that the positive rate of specific SARS antibody was high 10 days after disease onset. Testing specific antibody against SARS-CoV can help to confirm the diagnosis of SARS. However, it is necessary to further increase the sensitivity of early laboratory diagnosis of SARS.

Our results indicate that, while promoting the optimization and standardization of laboratory diagnostic reagents, one should combine several methods, such as RT-PCR, IFA and ELISA, for early and concise detection of SARS-CoV.

## References

1. Chinese Medical Association, China Association of Chinese Medicine, *Nat Med J China*, 2003, **83**: 1731–1752.
2. Xu GB *et al.*, *J Peking Univ*, 2003, **35**(Suppl): 23–25.
3. Fang LQ *et al.*, *Chin J Epidemiol*, 2003, **24**: 484.
4. Shi YL *et al.*, *Med J Chin People Liberal Army*, 2003, **28**: 572–574.
5. Lu EJ *et al.*, *Chin J Immunol*, 2003, **19**: 445–456.
6. Beijing Group of National SARS Prevention and Treatment of Emergency Technology, *Chin J Tuberc Respir Dis*, 2003, **26**: 339–342.

# 3.2

# Bacteriological and Molecular Identification of a Sporadic Case of Brucellosis in Shanghai

Cheng-Yan Meng, Fei-Yi Ruan, Jian-Qiang Lei,
Jia-Lin Jin and Wen-Hong Zhang*

*Huashan Hospital, Fudan University, Shanghai 200040, China*

**Objective.** To use bacteriological and molecular-biological methods to identify a sporadic case of brucellosis in Shanghai.

**Methods.** Blood culture combined with an automated biochemical system was used to isolate and identify a strain of *Brucella* from a patient suspected of brucellosis (a 54-year-old man who had frequent contact with live animals as a goat trader from Zhejiang province, which is not an endemic region of brucellosis in China; he was hospitalized for intermittent fever). A specific gene of *Brucella* was amplified from this strain and analyzed through sequencing. We employed the multiple-locus VNTR (variable number of tandem repeats) method to analyze the genotyping of this strain of *Brucella* by establishing a *Brucella* dendrogram tree using VNTR fingerprints based on published VNTR genotypes in the international database.

**Results.** The isolated strain from this sporadic case was identified to belong to the *Brucella* species by bacteriological and molecular-biological

*Corresponding author. Email: zhangwenhong@fudan.edu.cn

methods. Although biochemical analysis and specific gene sequencing could not identify the biovar of this species, a VNTR dendrogram of clustered VNTR genotypes indicated that this case of brucellosis belonged to *Brucella suis* biovar 2.

**Conclusion.** Conventional bacterial cultures and *Brucella* VNTR analysis can be used to identify *Brucella* infection and can investigate the molecular-epidemiological profile of strains isolated from infected patients.

*Keywords*: *Brucella*; VNTR.

## 1. Introduction

Brucellosis is a worldwide zoonotic disease caused by a number of host-adapted species of the Gram-negative bacterial genus *Brucella*. It is epidemic in the Mediterranean, the Middle East, West Asia, Africa and Latin America. In China, the highest prevalence of brucellosis is found in livestock regions, such as Inner Mongolia, or northwestern and northeastern regions of the country. As the disease typically affects individuals who work in contact with infected farm animals or with animal-derived tissues, brucellosis has been listed as an occupational infectious disease by the Ministry of Public Health. In Shanghai, patients who are infected with brucellosis are mainly those partaking in animal husbandry practices and those who come from epidemic regions. From 1957 to 1986, a total of 43 sporadic cases of brucellosis were discovered in Shanghai. The average annual incidence of brucellosis is estimated at approximately 0.03 out of 100 000 people. An epidemiological investigation from 1991 to 1992 into an occupational population in Shanghai reported that the incidence was 60 out of 100 000 people.[1] Recent increases in brucellosis cases have been dealt with in local Shanghai hospitals, and once a new infection is confirmed it is critical to determine, by epidemiological trace-back analysis, where the infection originated so as to prevent additional spread of the disease from this primary source. However, no such epidemiological study in China has been published. For this paper, a 54-year-old male who had frequent contact with live animals as a goat trader from Zhejiang province was hospitalized for intermittent fever. While he was in the hospital, two fluctuations in temperature, both between 38°C and 39°C, were observed during the course of one day. Apart from sweating after fever withdrawal, no other symptoms or signs were found. We tried to identify this suspected sporadic case of brucellosis in Shanghai by bacteriological and molecular-biological methods in addition to multiple-locus VNTR (variable number of tandem repeats) analysis, so as to identify the genotype of an isolated strain of *Brucella*. The patient was treated with streptomycin, rifampin and SMZ-TMP. From the second day of treatment, his symptoms improved and his temperature returned to normal. He completely recovered and was discharged 25 days later.

## 2. Results

### 2.1. *Microbiological tests*

Blood cultures were performed on two separate occasions; one showed *Pseudomonas stutzeri* and the other showed a Gram-negative bacillus. Brucellosis was finally considered given his occupational status and by ruling out tuberculosis or typhoid. Small, colorless and smooth colonies appeared on the blood agar plate after cultivation for 5–7 days and the isolated strain was further identified by an automated biochemical analysis system. Through microscopic analysis, it was determined that the colonies represented the typical sandlike distribution of *Brucella*; furthermore, these bacteria were determined to be aerobic, nonencapsulated, Gram-negative coccobacilli. Through analysis of a biochemical reaction of the strain (Table 1) and SAT, they were finally confirmed to be a *Brucella* species. Except for resistance to Aztreonam, the strain was susceptive to aminoglycosides, quinolones and cephalosporin.

### 2.2. *Results of BLAST*

The sequences of the PCR products amplified for the gene of a *Brucella*-specific secretion protein got three matches in NCBI BLAST. They were *Brucella melitensis* 1330, *Brucella abortus* 9-941 and *Brucella suis* 16M.

**Table 1.** Biochemical results on the isolated strain of *Brucella* from Shanghai, China.

| Biochemical reaction | Result |
|---|---|
| Oxidase | + |
| Urea enzyme | + |
| Nitrate reduction test | + |

**Table 2.** BLAST with published sequences of *Brucella*.

| Strain | Unmatched sequences |
|---|---|
| Sporadic strain (Shanghai, China) | GCCCGCCCTTTTTTGTATTGCCCTTGCTGTTGCCCTT GCTGTTGCCTATGCTGCCGCCAA |
| *Brucella abortus* (ref. NC_006932.1) | GCCCGCCCTTTTTTGTA ---------------------- TTGCCCT TGCTGTTGCCTATGCTGCCGCCAA |
| *Brucella melitensis* (ref. NC_003317.1) | GCCCGCCCTTTTTTGTA ---------------------- TTGCCCTT GCTGTTGCCTATGCTGCCGCCAA |
| *Brucella suis* (ref. NC_004310.3) | GCCCGCCCTTTTTTGTATTGCCCTTGCTGTTGCC- T ----------------------ATGCTGCCGCCAA |

However, there were 1% unmatched sequences (Table 2), which made it difficult to have additional subtyping.

## 2.3. *VNTR locus analysis*

Eight PCR products (150–200 bp) were obtained by using the HOOF-Print assay. After sequence analysis, the numbers of tandem repeats in each of the eight loci ranged from 1 to 11. Reference to the international *Brucella* VNTR database showed that genotypes were clustered into groups based on the differences in the numbers of repeat units at the eight VNTR loci (Fig. 1). In the dendrogram of clustered HOOF-Print genotypes, our isolated strain (labeled as an unknown strain) was in the *Brucella suis* evolutional tree and had the most recent genetic relationship with *Brucella suis* biovar 2.

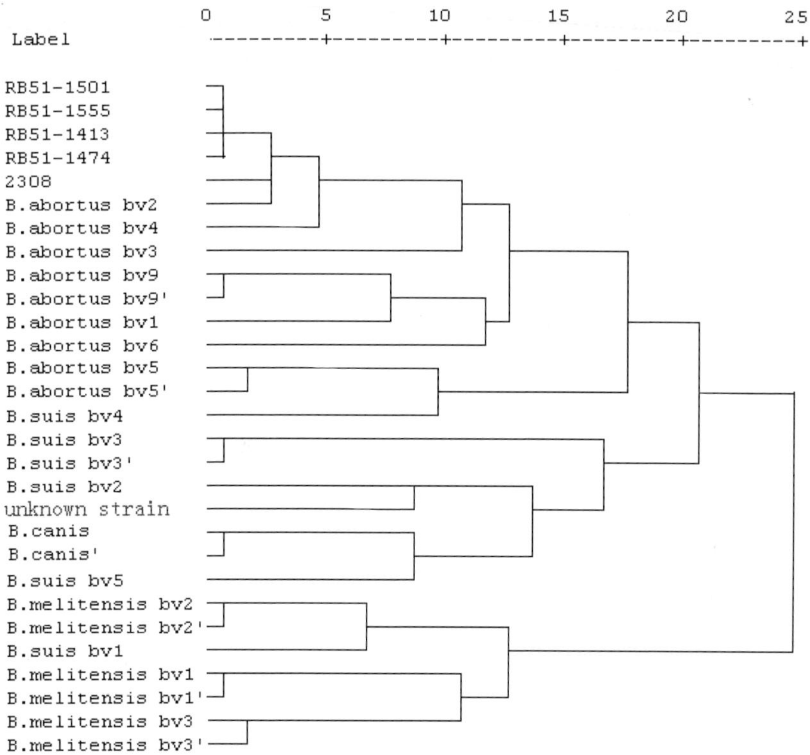


**Fig. 1.** *B. abortus*, *B. suis*, *B. melitensis* and *B. canis* are different species of *Brucella*. The strains labeled RB are vaccine strains and their parent strain 2308. A strain with an apostrophe means that there are two kinds of repeat numbers at the same locus.

## 3. Discussion

A patient with brucellosis was found in a low epidemic region in China with atypical clinical manifestations of the disease. Apart from fever and increased sweating, arthralgia, testicular pain, headache and neuralgia were absent in this particular patient. Although this case was once misdiagnosed as tuberculosis and typhoid, SAT, bacterial culture and biochemical identification finally helped us to make the correct diagnosis of brucellosis.

Currently, diagnosis of brucellosis mainly depends upon serologic methods which are designed to detect and measure antibodies to *Brucella* antigens.

The most frequently used test is SAT, but false-positive and false-negative results can occur as a result of immunologic cross-reactivity associated with many factors.[2,3] Therefore, a diagnosis of brucellosis should not be made based solely on SAT. The plate agglutination test (PAT) and the rose Bengal plate test (RBPT) are generally used as a means of preliminary screening, because these assays may produce non-specific agglutination. Certainly, new antibody tests, including the complement fixation test (CFT), which measures the IgG antibody, and ELISA, which measures both the IgM and IgG antibodies, are more sensitive and specific than SAT. They can be used for diagnosis of acute and chronic cases. The most conclusive means of establishing the diagnosis of brucellosis is by positive cultures from normally sterile body fluids or tissues. However, the rate of positive cultures is relatively low, due to harsh growth conditions and complex laboratory techniques. Bone marrow cultures are of higher yield than blood cultures. Blood may yield positive cultures in 7–14 days. However, even a positive culture is still difficult to be used for identification by most domestic automated biochemical techniques (we used the VITEK automated microbe analysis system). Recently, studies using immunoperoxidase staining and polymerase chain reaction (PCR) have been reported and these seem promising for purposes of diagnosis.[4]

In order to identify the isolated strain, we used PCR to amplify a specific *Brucella* gene for the secretion protein, and the products were further analyzed by sequencing. The PCR assay is a rapid and easy technique for detecting clinical samples directly. However, it could only identify the strain as one *Brucella* species and was unable to provide further information on this particular strain's genotype due to the genetic homogeneity within the genus.

Currently, there are very few tools available for subtyping *Brucella* isolates for epidemiological trace-back analysis. For this report, a PCR-based method was used to identify the number of repeat units at each locus, generating strain-specific fingerprints and establishing a dendrogram of clustered HOOF-Print genotypes of all published *Brucella* species and biovars. The technique successfully differentiated the strain as most likely *Brucella suis* biovar 2. In China, this report represents the first study on genotyping the *Brucella* species by multiple-locus analysis of

VNTRs. Although some *Brucella* species cannot be completely discriminated via analysis of a dendrogram, the analysis can be improved by increasing the sample size and accumulating more data. All in all, we are inclined to state that the technique presented in this paper is highly discriminatory against *Brucella* species among unrelated field isolates. This technology is especially useful in the epidemiological investigation of an outbreak, providing insight into the evolution of a brucellosis outbreak and methods for preventing further spread of the disease.[5] We are aware that a large-scale research on the epidemiology of brucellosis in China is needed in order to evaluate the HOOF-Print assay employed in this study.

## References

1. Shi HY *et al.*, *Chin J Ind Med*, 1994, **7**(5): 291–292.
2. Yang T *et al.*, *Chin J Clin*, 2005, **33**(6): 29–30.
3. Ding LP *et al.*, *Clin Misdiag Misther*, 2002, **15**(5): 355–356.
4. Qi JW *et al.*, *Chin J Veter Med*, 2004, **40**(9): 50–52.
5. Bricker BJ *et al.*, *BMC Microbiol*, 2005, **5**: 37.

# 3.3

# A PCR-Capillary Electrophoresis Method Detects and Differentiates Biovars of *Ureaplasma urealyticum*

Shi-Hang Zhou* and Zhuo-Ran Zhang[†,‡]

**Dalian Red Cross Blood Center, Dalian 116001, China*
[†]*Department of Microbiology, Dalian Medical University, Dalian 116044, China*

**Objective.** To develop a polymerase chain reaction–capillary electrophoresis (PCR–CE)-based method for detection and differentiation of biovar 1 and biovar 2 of *Ureaplasma urealyticum* and to determine whether *U. urealyticum* biovars are associated with nongonococcal urethritis (NGU) in male patients.

**Methods.** Primers for two biovars of *U. urealyticum* were synthesized for PCR amplification, and PCR products were detected and differentiated using CE. Two biovars of *U. urealyticum* were detected by the above method in the urine samples from different male populations. The sensitivity and specificity of this PCR–CE-based method were determined.

**Results.** This method showed an analytic sensitivity of 10 copies/50 μl, with both biovar 1 and biovar 2 of *U. urealyticum* specifically amplified.

[‡]Corresponding author. Email: zzr1941577@163.com

The prevalence of *U. urealyticum* biovar 2 in the NGU group and the nonchlamydial NGU (NCNGU) group was higher than that in the control group ($P < 0.05$), but the prevalence of *U. urealyticum* biovar 1 in the NGU group, the NCNGU group and the control group was not significantly different ($P > 0.05$).

**Conclusion.** A PCR–CE-based method possesses good sensitivity and specificity for the detection and differentiation of two biovars of *U. urealyticum*. *U. urealyticum* biovar 2, not biovar 1, may be associated with NGU in males.

*Keywords*: *Ureaplasma urealyticum*; biovar; polymerase chain reaction; capillary electrophoresis; nongonococcal urethritis.

## 1. Introduction

*Ureaplasma urealyticum* is a member of the genus *Ureaplasma*, the family *Mycoplasmataceae*, the order *Mycoplasmatales* and the class *Mollicutes*, and is divided into two biovars: biovar 1 and biovar 2. Biovar 1 is composed of serotypes 1, 3, 6 and 14, whereas biovar 2 is composed of 10 serotypes, which are numbered 2, 4, 5, 7, 8, 9, 10, 11, 12 and 13.[1] *U. urealyticum* can cause a number of human diseases. Generally, it was considered that *U. urealyticum* was associated with nongonococcal urethritis (NGU). Recently, many studies have shown that *U. urealyticum* is not associated with NGU.[2–4] In 1999, biovar 1 was designated as a separate species, *Ureaplasma parvum*, whereas biovar 2 retained the *U. urealyticum* designation.[5] Thus, an association between *U. urealyticum* and NGU should be studied in terms of biovars.

In the present study, we developed a molecular method for detection and differentiation of two biovars of *U. urealyticum* in a combination of PCR amplification and capillary electrophoresis detection (PCR–CE). Two biovars of *U. urealyticum* were detected in urine samples from different male populations using this method in order to determine whether the two biovars are associated with NGU in males.

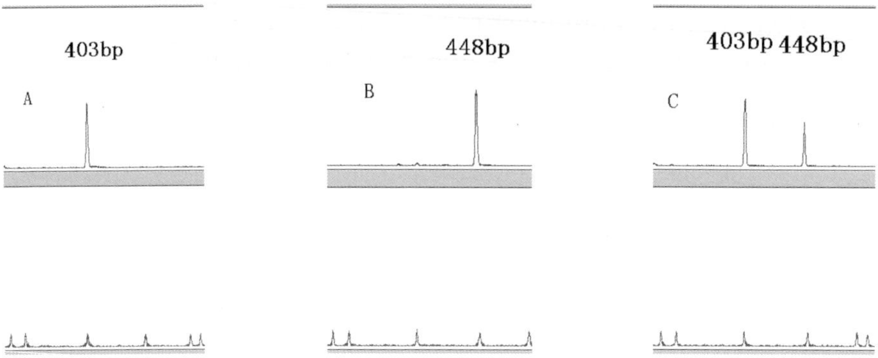


**Fig. 1.** Capillary electrophoresis of PCR products obtained with reference strains of *U. urealyticum.* (A) *U. urealyticum* biovar 1; (B) *U. urealyticum* biovar 2; (C) *U. urealyticum* biovar 1 and biovar 2.

## 2. Results

### 2.1. *PCR–CE of reference strains of* U. urealyticum

As expected, the length of the PCR product obtained with *U. urealyticum* biovar 1 was 403 bp. However, for *U. urealyticum* biovar 2 it was 448 bp (Fig. 1).

### 2.2. *Analysis of sensitivity*

Ten copies of *U. urealyticum* were detected using PCR–CE, but one copy was below the detection limit and so could not be detected. Thus, the analytic sensitivity of PCR–CE was 10 copies/50 μl. In addition, the analytic sensitivity of PCR–CE for the mixed sample was the same.

### 2.3. *Analysis of specificity*

This PCR–CE only amplified *U. urealyticum* biovar 1 and biovar 2 specifically and had no cross-reaction with bacteria, viruses or fungi used in the study.

## 2.4. *Detection of* N. gonorrhoeae, C. trachomatis, U. urealyticum *biovar 1 and biovar 2 in men*

*N. gonorrhoeae*, *C. trachomatis*, *U. urealyticum* biovar 1 and biovar 2 detected in 103 men are summarized in Table 1. Thirty-four out of 70 men (48.6%) with urethritis were positive for *N. gonorrhoeae*. Of those men with gonococcal urethritis, 3 (8.8%) were also positive for *U. urealyticum* biovar 1, and 8 (23.5%) were also positive for *U. urealyticum* biovar 2. Of the remaining 36 (51.4%) men with NGU, 5 (13.9%) were positive for *C. trachomatis*, 6 (16.7%) were positive for *U. urealyticum* biovar 1 and 13 (36.1%) were positive for *U. urealyticum* biovar 2. Of 33 men in the control group, 10 (30.3%) were positive for *U. urealyticum* biovar 1, and 2 (6.1%) were positive for *U. urealyticum* biovar 2. In addition, 1 (3.0%) was positive for *U. urealyticum* biovar 1 and biovar 2.

The prevalence of *U. urealyticum* and its biovars in the urethritis group and in the control group are shown in Table 2. The prevalence of *U. urealyticum* biovar 1 in the control group was significantly higher than that in the urethritis group ($P < 0.05$), and the prevalence of *U. urealyticum* biovar 2 in the urethritis group was significantly higher

**Table 1.** *N. gonorrhoeae*, *C. trachomatis*, *U. urealyticum* biovar 1 and biovar 2 detected in men.

| | Case number | *N. gonorrhoeae* | *C. trachomatis* | *U. urealyticum* biovar 1 | *U. urealyticum* biovar 2 |
|---|---|---|---|---|---|
| Gonococcal | 3 | + | − | + | − |
| urethritis | 8 | + | − | − | + |
| group ($n = 34$) | 23 | + | − | − | − |
| NGU group | 2 | − | + | + | − |
| ($n = 36$) | 3 | − | + | − | + |
| | 4 | − | − | + | − |
| | 10 | − | − | − | + |
| | 17 | − | − | − | − |
| Control group | 10 | − | − | + | − |
| ($n = 33$) | 1 | − | − | + | + |
| | 2 | − | − | − | + |
| | 20 | − | − | − | − |

**Table 2.** Prevalence of *U. urealyticum* and its biovars in the urethritis group and the control group.

| | Urethritis group ($n$ = 70) | Control group ($n$ = 33) | $P$ value |
|---|---|---|---|
| | Case number (%) | Case number (%) | |
| *U. urealyticum* biovar 1 | 9 (12.9) | 11 (33.3) | 0.030 |
| *U. urealyticum* biovar 2 | 21 (30.0) | 3 (9.1) | 0.024 |
| *U. urealyticum** | 30 (42.9) | 13† (39.4) | 0.832 |

**U. urealyticum* is positive when *U. urealyticum* biovar 1 or biovar 2 is detected.
†Both *U. urealyticum* biovar 1 and biovar 2 from one case in the control group are positive.

than that in the control group ($P$ < 0.05). But the prevalence of *U. urealyticum* in the urethritis group did not differ significantly from that in the control group ($P$ > 0.05).

The prevalence of *U. urealyticum* and its biovars in different urethritis groups and in the control group are shown in Table 3. The prevalence of *U. urealyticum* biovar 1 in the control group was significantly higher than that in the gonococcal urethritis group ($P$ < 0.05), but the prevalence of *U. urealyticum* and *U. urealyticum* biovar 2 in the gonococcal urethritis group did not differ significantly from that in the control group ($P$ > 0.05). The prevalence of *U. urealyticum* biovar 2 in the NGU group was significantly higher than that in the control group ($P$ < 0.05), but the prevalence of *U. urealyticum* and *U. urealyticum* biovar 1 in the NGU group did not differ significantly from that in the control group ($P$ > 0.05). The prevalence of *U. urealyticum* biovar 2 in the nonchlamydial NGU (NCNGU) group was significantly higher than that in the control group ($P$ < 0.05), but the prevalence of *U. urealyticum* and *U. urealyticum* biovar 1 in the NCNGU group did not differ significantly from that in the control group ($P$ > 0.05).

## 3. Discussion

Culture is currently the most widely used method for detection of *U. urealyticum* in clinical laboratories. It is simple, but it is time-consuming and cannot distinguish between two biovars of *U. urealyticum*. At present,

**Table 3.** Prevalence of *U. urealyticum* and its biovars in different urethritis groups and the control group.

| | Gonococcal urethritis group ($n = 34$) | NGU group ($n = 36$) | Nonchlamydial NGU group ($n = 31$) | Control group ($n = 33$) | $P$ value |
|---|---|---|---|---|---|
| | Case number (%) | Case number (%) | Case number (%) | Case number (%) | |
| *U. urealyticum* biovar 1 | 3 (8.8) | 6 (16.7) | 4 (12.9) | 11 (33.3) | 0.017*<br>0.162†<br>0.077‡ |
| *U. urealyticum* biovar 2 | 8 (23.5) | 13 (36.1) | 10 (32.3) | 3 (9.1) | 0.186*<br>0.010†<br>0.030‡ |
| *U. urealyticum*[a] | 11 (32.3) | 19 (52.8) | 14 (45.2) | 13[b] (39.4) | 0.615*<br>0.336†<br>0.801‡ |

[a] *U. urealyticum* is positive when *U. urealyticum* biovar 1 or biovar 2 is detected.
[b] Both *U. urealyticum* biovar 1 and biovar 2 from one case in the control group are positive.
*Gonococcal urethritis group vs. control group.
†NGU group vs. control group.
‡ Nonchlamydial NGU group vs. control group.

many methods, including conventional PCR,[1] PCR–SSCP,[6] PCR- and phylogeny-based assay,[7] PCR–microtiter plate hybridization assay,[8] reverse line blot hybridization[9] and PCR–liquid hybridization,[10] for detection of two biovars of *U. urealyticum* have been reported based on the difference of the MB antigen gene, 16S rRNA gene, 16S–23S rRNA gene and urease gene between *U. urealyticum* biovar 1 and biovar 2. Among these methods, the PCR developed by Teng *et al.*[1] is a relatively simple and rapid one. Because the products from biovar 2 were approximately 45 bp longer (448 bp) than the 403 bp biovar 1 fragment, the low resolution of agarose gel electrophoresis might result in error. To overcome the disadvantage, we have developed PCR–CE using CE. Distinguishing between two biovars of *U. urealyticum* using CE is satisfactory. The results of our study indicated that the sensitivity of PCR–CE is 10 copies/50 μl, and that PCR–CE only amplified *U. urealyticum* biovar 1 and biovar 2 specifically and had no cross-reaction with bacteria and fungi often found in the urethra and the hepatitis B virus. Teng *et al.*[1] found that this PCR had no cross-reaction with 17 kinds of mycoplasma whose phylogeny is close to *U. urealyticum*. This PCR–CE is suitable for use in the clinical laboratory because of its high resolution, strong specificity and high sensitivity.

In this study we found that the prevalence of *U. urealyticum* biovar 2 in the urethritis group, NGU group and NCNGU group was significantly higher than that in the control group, suggesting that *U. urealyticum* biovar 2 was associated with NGU in males, and that it might play a pathogenic role in NGU. In addition, we found that the prevalence of *U. urealyticum* biovar 1 in the control group was significantly higher than that in the urethritis group and gonococcal urethritis group, but the prevalence of *U. urealyticum* biovar 1 in the NCNGU group and NGU group did not differ significantly from that in the control group. These results indicated that in the subjects we studied, *U. urealyticum* biovar 1 was not associated with NGU in males and might only be the result of colonization of *U. urealyticum*. These results could also explain why *U. ureaplasma* is frequently recovered from the urethra of asymptomatic men.

Our study also showed that if two biovars of *U. ureaplasma* were not distinguished, there would be no significant difference in the prevalence of *U. ureaplasma* between different urethritis groups and the control group. This result is consistent with other studies.[2–4,11] Deguchi *et al.*[11] indicated that *U. urealyticum* biovar 2 was associated with NGU, and that *U. urealyticum* biovar 1 in the male urethra might be the result of colonization. Though our results are consistent with theirs, the prevalence of *U. ureaplasma* and its biovars in the populations of our study was higher than that of theirs.

## References

1. Teng LJ *et al.*, *J Clin Microbiol*, 1994, **32**: 1464–1469.
2. Jiang J *et al.*, *China J Lepr Skin Dis*, 2005, **21**: 85–88.
3. Morency P *et al.*, *Sex Transm Infect*, 2001, **77**: 125–129.
4. Totten PA *et al.*, *J Infect Dis*, 2001, **183**: 269–276.
5. Kong F *et al.*, *Int J Syst Bacteriol*, 1999, **49**: 1879–1889.

6. Pitcher D *et al.*, *J Clin Microbiol*, 2001, **39**: 1840–1844.
7. Yoshida T *et al.*, *J Clin Microbiol*, 2002, **40**: 105–110.
8. Yoshida T *et al.*, *J Clin Microbiol*, 2003, **41**: 1850–1855.
9. Wang H *et al.*, *Appl Environ Microbiol*, 2004, **70**: 1483–1486.
10. Povlsen K *et al.*, *J Clin Microbiol*, 1998, **36**: 3211–3216.
11. Deguchi T *et al.*, *Sex Transm Dis*, 2004, **31**: 192–195.

# 3.4

# Implications of CD64 in Early Diagnosis of Neonatal Septic Cases

Rong Zhang and Chao Chen*

*Department of Neonatology, Children's Hospital of Fudan University, Shanghai 201102, China*

**Objective.** To assess the level of the leukocyte surface antigen CD64 during neonatal septicemia, and to assess the implications of CD64 in early diagnosis of neonatal septicemia.

**Methods.** Forty-eight newborns were classified into three groups: septicemia, local infection, and disease without infection. Peripheral blood samples of CD64 and CRP from these patients were compared.

**Results.** The levels of CD64 in the septicemia group were significantly higher than those observed in the other two groups ($P < 0.001$). According to the ROC curve, when CD64 was $\geq$ 4504 molecules/unit, the sensitivity for identifying neonatal septicemia was 90.1% and the specificity was 100%.

**Conclusion.** The level of CD64 was increased during neonatal septicemia. CD64 could be an effective diagnosis marker in neonatal septicemia.

*Keywords*: CD64; newborn; septicemia; infection.

*Corresponding author. Email: chen6010@163.com

# 1. Introduction

CD64 (Fcγ-RI) is expressed at a low level on the surface of nonactivated neutrophils in healthy subjects. However, during bacterial infection, its expression will be substantially upregulated, leading to neutrophil activation.[1] It has been reported that for adults and children, the upregulation of CD64 may be useful for early diagnosis of bacterial infection and evaluation of the severity of disease. In the present study, we investigated the expression of CD64 in bacteria-infected newborns, especially those with septicemia, and estimated its application as a potential diagnostic marker for septicemia.

# 2. Results

## 2.1. *The expression of CD64*

A total of 48 newborns were studied. The expression profiles of the three groups are listed in Table 1. Statistical analysis showed that the expression of CD64 among these three groups was significantly different. The expression of CD64 from the septicemia group was higher than that from the local infection group ($P < 0.001$). In addition, the expression of CD64 in blood from the local infection group was also higher than that from the noninfection group ($P < 0.001$).

## 2.2. *The levels of CRP*

Results showed that the average level of CRP in the septicemia group was higher than that in the local infection and noninfection groups. Statistical

**Table 1.** Serum levels of CD64 expression (molecules/cell).

| Group | *n* | Mean ± SD |
|---|---|---|
| Septicemia | 21 | $8255.5 \pm 3641.7^{1}$ |
| Local infection | 14 | $2944.1 \pm 727.9^{2}$ |
| Noninfection | 13 | $1702.6 \pm 559.9^{3}$ |

$^{1}P < 0.001$ between septicemia group and local infection group; $^{2}P < 0.001$ between local infection group and noninfection group; $^{3}P < 0.001$ between septicemia group and noninfection group.

analysis showed that the CRP level in the septicemia group was higher than that in noninfection group ($P < 0.001$), and the CRP level in the septicemia group was also higher than those in the local infection group ($P < 0.001$) (Table 2). However, there was no significant difference in CRP levels between the local infection and noninfection groups ($P = 0.076$).

### 2.3. *The sensitivity and specificity of CD64 and CRP in the diagnosis of neonatal septicemia*

The ROC curve was determined by combining the data of the septicemia group, local infection group and noninfection group. As shown in Table 3 and Fig. 1, when CD64 reached 2964 molecules/cell, it showed a sensitivity of 100% and a specificity of 77.8% for the diagnosis of septicemia. When the cutoff value was set as 4505 molecules/cell, CD64 showed a higher sensitivity of 90.5% and a specificity of 100%. In contrast, the cut-off value of CRP at 9 provided only a sensitivity and a specificity of 81% and 77.8%, respectively. In addition, when the cutoff value of CRP was

**Table 2.** Serum CRP concentrations (mg/L).

| Group | *n* | Mean ± SD |
|---|---|---|
| Septicemia | 21 | 2137.9 ± 26.6 |
| Local infection | 14 | 12.7 ± 7.1[1] |
| Noninfection | 13 | 7.6 ± 1.4[2] |

[1] $P = 0.022$ between septicemia group and local infection group; [2] $P < 0.001$ between septicemia group and noninfection group.

**Table 3.** Comparison of sensitivity and specificity of markers using different cutoff values.

| Infection marker | Cutoff value | Sensitivity (%) | Specificity (%) |
|---|---|---|---|
| CD64 (molecules/cell) | 2964 | 100 | 77.8 |
| | 4505 | 90.5 | 100 |
| CRP (mg/L) | 9 | 81 | 77.8 |
| | 16 | 57.1 | 85.2 |

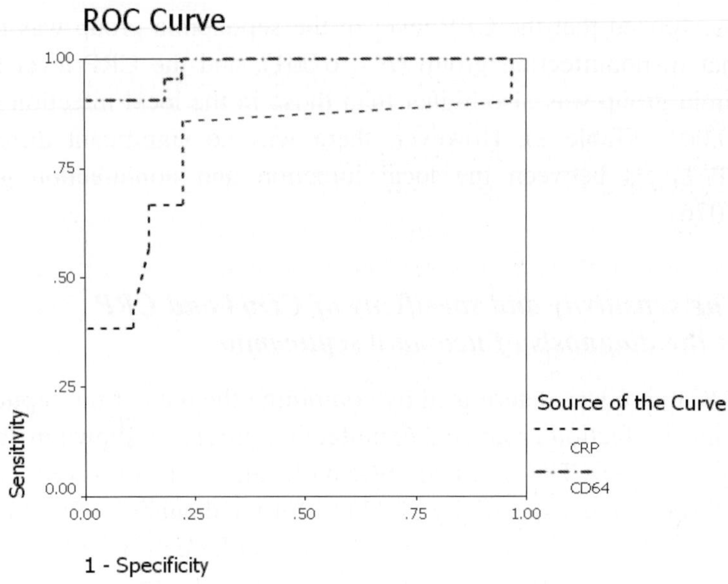


**Fig. 1.** The ROC curves of CD64 and CRP.

set at 16, the sensitivity was decreased to 57.1%, and the specificity was increased to 85.2%.

## 2.4. *Dynamic changes of CD64 during septicemia*

Patients who were diagnosed as sepsis were retested for the expression of CD64 in the course of 5–7 days, following admission (excluding those who died or were discharged during this period) ($n = 12$). CD64 levels at the time of admission were compared, and the expression levels of CD64 significantly decreased, except for two cases ($P < 0.001$). For the two cases that did not show a decrease in CD64 levels, recurrent infections and a worsening condition were observed. In addition, one patient died three days after admission and three cases were transferred to the Neonatal Intensive Care Unit (NICU) as a result of critical levels of CD64 (>13 000 molecules/cell). The conditions of these three patients improved after antimicrobial therapy and symptomatic treatment, and CD64 expression levels decreased significantly.

## 3. Discussion

Infection is still one of the most common diseases in neonates. The morbidity and mortality of severe infections, such as septicemia, are still high. Early diagnosis and treatment are essential for prognosis. Commonly used markers such as white blood cell count and CRP are not sensitive enough for early diagnosis. Furthermore, the positive rate of blood culture, which is a gold standard for diagnosis, is low. Several studies have demonstrated that interleukin 6 (IL-6), tumor necrosis factor α (TNF-α)[2] and procalcitonin (PCT)[3] can be early diagnosis markers of newborn septicemia. However, these cytokines reach a peak in the early duration of infection and decrease rapidly. Therefore, their sensitivity and specificity for early diagnosis of septicemia depend on the time when the blood sample is collected. In this study, we aimed to determine the diagnostic use of expression of the neutrophils, CD64, which reflects host immune response to bacterial infection.

Neutrophils play an important role in host nonspecific responses to bacterial infections. To be an effective antibody or complement, the binding to receptors on the cell membrane (such as complement receptor Fcγ-R) is required. Three major classes of Fcγ-R, including Fcγ-RI (CD64), Fcγ-RII (CD32) and Fcγ-RIII (CD16), have been characterized on neutrophils and monocytes. CD64 is usually expressed at a low level on the surface of nonactivated neutrophils and upregulated during a bacterial infection. The ligation between CD64 and immunoglobulin Fc fragments triggers a number of important biological functions, such as respiratory burst, and antibody-dependent cytotoxicity, resulting in activation of antibacterial immunity.[1] Humoral immunity in newborns, especially premature infants, is usually low, while cellular immunity, including the function of neutrophils, plays a critical role in the antibacterial immune response.

It has been demonstrated that expression of the neutrophil CD64 is significantly upregulated in adults during a bacterial infection,[4,5] which reflects the severity of the disease and prognosis. Barth *et al.* found that neutrophils from children with septic shock expressed a high level of CD64 which decreased rapidly after treatment.[6] Our findings showed that neutrophils from newborns with septicemia also expressed higher levels of CD64 compared to those with local infection or no infection, which suggested that CD64 is also a sensitive marker for newborn septicemia, similar to the observation by Ng *et al.*[7,8] In our study, when the cutoff value for CD64 was 4505 molecules/cell, CD64 showed a high sensitivity of 90.5% and a specificity of 100%, indicating that CD64 is a sensitive diagnostic marker for the identification of early septicemia. In addition, based on the results of followup, the expression level of CD64 was shown to relate to the progress of disease.

CRP, an acute reactive protein, is a widely used indicator for infection. In our study, different CRP values were observed among the three groups. Although CRP was effective as a marker for severity of infection, it increased slowly after infection and was of relatively low sensitivity in early diagnosis of infection. Also, it was not increased in a few cases of septicemia, which might be attributed to the blood sample being taken before the peak period of infection. In comparing CD64 with CRP, the diagnosis of infection was consistent, while CD64 showed a higher sensitivity and specificity in early diagnosis of neonatal septicemia.

In conclusion, our findings suggest that the neutrophil CD64 is a potential marker for early diagnosis of neonatal septicemia. When symptoms and signs suggest neonatal septicemia, and if CD64 > 2964 molecules/cell, neonatal septicemia should be considered. If CD64 > 4504 molecules/cell, the chances of neonatal septicemia would be extremely high. Up until this point, no neutrophil CD64 measurement has been taken in neonates. More data should be collected to set the cutoff value for its implication in septicemia diagnosis.

## References


1. Fjaertoft G *et al.*, *Pediatr Res*, 1999, **45**(6): 871–876.
2. Silveira R *et al.*, *Acta Paediatr*, 1999, **88**: 647–650.
3. Casado Flores J *et al.*, *An Esp Pediatr*, 2001, **54**: 69–73.
4. Herra CM *et al.*, *J Med Microbiol*, 1996, **44**: 135–140.
5. Leino L *et al.*, *Clin Exp Immunol*, 1997, **107**: 37–43.
6. Barth E *et al.*, *Cytokine*, 2001, **14**(5): 299–302.
7. Ng PC *et al.*, *Pediatr Res*, 2002, **51**(3): 296–303.
8. Ng PC *et al.*, *Pediatr Res*, 2004, **56**(5): 796–803.

# 3.5

# Correlation Between the MICA*A5.1 Allele and *Chlamydia trachomatis* Infection

Bing Mei, Kun Du, Zhi Huo, Fu-Yan Wang, Shu-Hui Ma and Ping Yu*

*Department of Immunology, Xiangya Medical School of Central South University, Changsha 410008, China*

**Objective.** To investigate whether there is correlation between an individual's genetic background and *Chlamydia trachomatis* (*Ct*) infection through analysis of the MICA*A5.1 allele and *Ct* infection in the infertile and control groups.

**Methods.** The MICA*A5.1 allele was analyzed by PCR-SSP, and FQ-PCR tested *Ct* DNA.

**Results.** Out of 122 cases, 30 cases (27.1%) were *Ct*-positive in the infertile group, and out of 140 individuals, 16 (11.4%) were *Ct*-positive in the control group. Among infertile women, 35 cases were MICA*A5.1 allele-positive. In the control group, there were 43 people whose MICA*A5.1 allele was positive. In the infertile group, 5 patients infected with *Ct* were MICA*A5.1 allele-positive, whereas only 1 case had the same results in the control group. The difference in *Ct* infection rates between MICA*A5.1-positive cases and MICA*A5.1-negative cases was

*Corresponding author. Email: yuping1953@sina.com

statistically significant ($P = 0.046$ in the infertile group; $P = 0.022$ in the control group; $P = 0.01$ in infertile women and control pooled together).

**Conclusion.** *Ct* infection rates are higher in infertile women than in controls, and individuals with a positive MICA*A5.1 allele may have lower susceptibility to *Ct* infection than ones with a negative allele.

*Keywords*: MHC class I chain-related gene A; *Chlamydia trachomatis*; infection.

## 1. Introduction

*Chlamydia trachomatis* (*Ct*) is an obligate intracellular bacterium with a high infection rate in the population. It is also a major cause of trachoma, sexually transmitted diseases, and infertility. In addition, *Ct* infection may correlate with cervical carcinoma.[1] Major histocompatibility complex (MHC) class I chain-related gene A (MICA) is found to be a functional gene with immunological significance. The protein encoded by MICA can activate NK cells, $CD8^{+}T$ cells and $\gamma\delta T$ cells. The MICA transmembrane region includes a microsatellite polymorphism consisting of the repetition of a trinucleotide (GCT), situated in exon 5. The alleles MICA*A4, A5, A6 and A9 consist of variable numbers of GCT repeats, encoding 4, 5, 6 or 9 alanine residues, respectively. The MICA*A5.1 allele, which encodes a protein with the short transmembrane and cytoplasm region, is the result of frameshift mutation of allele 5 with one guanine insertion. Herein, we investigate the MICA*A5.1 allele and *Ct* infection in normal controls and infertile patients to reveal whether there is a correlation between the MICA*A5.1 allele and *Ct* susceptibility.

## 2. Results

### 2.1. *MICA gel electrophoresis map*

The two pairs of primers were designed according to the sequences reported. The second pair of primers was designed only to amplify MICA*A5.1, while the first pair of primers amplified any MICA allele.

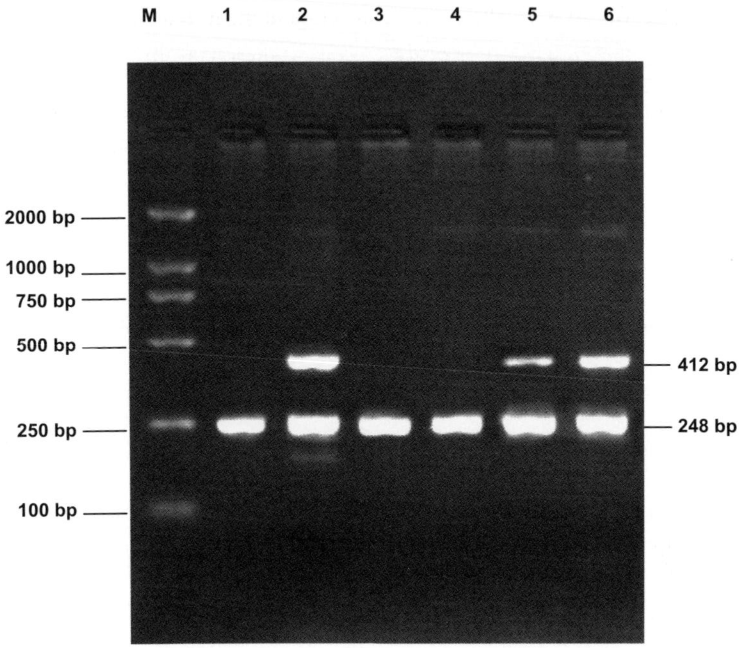


**Fig. 1.** Gel electrophoresis of PCR-SSP products to detect the MICA*A5.1 allele. Lane 1 – Negative control, MICA*A5.1 (–), DNA from the U373 cell. Lane 2 – Positive control, MICA*A5.1 (+), DNA from the HeLa cell. Lanes 3 and 4 – Test sample, MICA*A5.1 (–). Lanes 5 and 6 – Test sample, MICA*A5.1 (+).

The control DNA allele was obtained from homozygote, HeLa and U373 cell lines. HeLa is a MICA*5.1(+) homozygote, while the U373 cell is a MICA*5.1(–) homozygote (Fig. 1).

## 2.2. *The MICA*A5.1 allele and* Ct *infection in the infertile and control groups*

In the infertile group ($n = 122$), 33 cases were determined to be *Ct*-positive. In the control group ($n = 140$), 16 patients were *Ct*-positive. The *Ct* infection rate of the infertile group was higher than that of the control group (Table 1).

**Table 1.** The MICA*A5.1 allele and *Ct* infection in the infertile and control groups.

| Group | MICA*A5.1 | | *Ct* infection | | *Ct* infection rate |
|---|---|---|---|---|---|
| | + | − | + | − | (%) |
| Infertile women | 35 | 87 | 33 | 89 | 27.1 |
| Control women | 43 | 97 | 16 | 124 | 11.4 |

**Table 2.** Correlation between *Ct* infection and the MICA allele.

| Group | *Ct* infection | | *Ct* infection rate |
|---|---|---|---|
| | + | − | (%) |
| Sterile patients | | | |
| MICA*A5.1 (+) | 5 | 30 | 14. 3 |
| MICA*A5.1 (−) | 28 | 59 | 32. 2 |
| Controls | | | |
| MICA*A5.1 (+) | 1 | 42 | 2.3 |
| MICA*A5.1 (−) | 15 | 82 | 15.5 |
| Sterile patients and control pool | | | |
| MICA*A5.1 (+) | 6 | 72 | 7.7 |
| MICA*A5.1 (−) | 43 | 141 | 23.4 |

## 2.3. *Correlation between* Ct *infection and the MICA allele in the infertile and control groups*

Among the infertile women ($n$ = 122), 35 cases were MICA*A5.1 allele-positive, while in the control group ($n$ = 140), 43 patients were so. In the infertile group, 5 patients infected with *Ct* were MICA*A5.1 allele-positive, whereas only 1 patient was *Ct*-positive and MICA*A5.1 allele-positive in the control group. The difference in *Ct* infection rates between MICA*A5.1-positive cases and MICA*A5.1-negative cases was statistically significant ($P = 0.046$ in the infertile group; $P = 0.022$ in the control group *and* $P = 0.01$ in infertile women and control pooled together) (Table 2).

## 3. Discussion

*Ct* is an obligate intracellular bacterium with a high infection rate in the population and can cause many kinds of diseases. Barlow *et al.* had proven that *Ct* DNA existed in the uterine tube and uterus of infertile patients through PCR and *in situ* hybridization methods.[2] In our study, using *Ct* DNA samples from the cervix or uterus, the *Ct* infection rate (27.1%) in the infertile group was higher than that (11.4%) in the control group, implying that *Ct* infection may be the cause of infertility. After *Ct* infection, high levels of CHSP60 were produced in the uterine tube, which could break the balance of Th1 and Th2 responses and lead to abnormal cellular and humoral immunity. *Ct* is usually not effectively cleared by the host and results in persistent infection. With the persistent infection, adherence and blockade can be formed in the uterine tube.[3] At the same time, CHSP60 of *Ct* has a higher homology with that from the human and could cause persistent damage to host tissues through the induction of autoimmunity.[4]

MICA is in close proximity to the HLA-B and HLA-C loci. The MICA transmembrane region includes a microsatellite polymorphism consisting of a repetition of a trinucleotide (GCT), situated in exon 5. The alleles MICA*A4, A5, A6 and A9 consist of variable numbers of GCT repeats, encoding 4, 5, 6 or 9 alanine residues, respectively. The MICA*A5.1 allele, which encodes protein transmembrane and cytoplasm amino acid residues, harbors a frameshift mutation of allele 5 with one guanine insertion. The protein encoded by MICA can activate NK, $CD8^{+}T$ and γδT cells.[5] From the study of MICA5.1 and *Ct*, it is interesting to find that individuals with a positive MICA*A5.1 allele had lower susceptibility to *Ct* infection than patients with a negative allele in both the control and infertile groups. It has been suggested that an individual's genetic background may affect their susceptibility to *Ct* infection. Currently, the mechanisms of this phenomenon are unknown, but there are some reports on changes in MICA after bacterial or viral infections. For example, it has been found that binding of *Escherichia coli* adhesin AfaE to CD55 triggered cell surface expression of MICA.[6] *In vitro*, full-length alleles of MICA were eliminated from the cell surface after infection with human cytomegalovirus (HCMV). When the MICA*5.1 allele was found in infected host cells, cells could be cytolyzed by NK cells, while cells of other alleles could not.[7] In recent experiments it was shown that in polarized intestinal cells, most MICA molecules were located on the basolateral surface, while MICA*5.1 was transported to the apical region of the surface. Results suggested that MICA*5.1 could be transported by a different pathway from the full-length alleles, which could protect the molecules encoded by MICA*5.1 from degradation in cells.[8] We speculate that after *Ct* infection, MICA protein encoded by the MICA*5.1 allele could be highly expressed on cell surfaces, and may activate NK, $CD8^{+}T$ and $\gamma\delta T$ cells to kill *Ct*-infected cells, thereby protecting the host from persistent *Ct* infection. These speculations merit further experimental studies.

## References


1. Wallin KL *et al.*, *Inter J Cancer*, 2002, **101**: 371–374.
2. Barlow RE *et al.*, *J Med Microbiol*, 2001, **50**: 902–908.
3. Kinnunen A *et al.*, *Clin Exp Immunol*, 2003, **131**: 299–303.
4. LaVerda D *et al.*, *Infect Dis Obstet Gynecol*, 1999, **7**: 64–71.
5. Collins RW *et al.*, *Eur J Immunogenet*, 2004, **31**: 105–114.
6. Tieng V *et al.*, *Proc Natl Acad Sci USA*, 2002, **99**: 2977–2982.
7. Zou Y *et al.*, *J Immunol*, 2005, **174**: 3098–3104.
8. Suemizu H *et al.*, *Proc Natl Acad Sci USA*, 2002, **99**: 2971–2976.

# 3.6

# Surveillance of Human Infections with Avian Influenza in Guangzhou During 1997–2006

Peng-Zhe Qin, Yu-Lin Wang, Biao Di, Xiu-Zhen Zhou, Yang Liu, Yan-Shan Cai, Ji-Chuan Shen, Yu-Fei Liu, En-Jie Lu, Duan-Hua Zhou and Ming Wang*

*Guangzhou Center for Disease Control and Prevention, Guangzhou 510080, China*

**Objective.** To understand the epidemiological features of human infection with avian influenza in a special population in Guangzhou.

**Methods.** Surveillance was conducted in Guangzhou from 1997 among a group of workers involved in breeding, selling and slaughtering of poultry and other animals which may potentially carry avian influenza viruses. Serum samples were collected from this high risk population to detect specific antibodies via the microhemagglutination inhibition test. An epidemiological survey was conducted on people infected with H9N2 or H5N1.

**Results.** The H9 infection rate among the occupational group that handled poultry in Guangzhou was 4.56% (0.51%–10.00%). There was no significant difference among different age groups. One characterized

*Corresponding author. Email: wangming@gzcdc.org.cn

human case of avian influenza A (H5N1) in Guangzhou in 2006 was reported to have visited many live poultry food markets before the onset of the disease. A poultry slaughterer working in one of these live poultry food markets had a positive anti-H5N1 antibody response, while a person with anti-H5N1 antibody and a patient with anti-H9N2 antibody had no history of direct contact with poultry.

**Conclusion.** H5N1 and H9N2 may cause an asymptomatic or low symptomatic infection in humans. In addition to people working in poultry farms, those involved in avian breeding, selling and slaughtering are also at high risk for avian influenza viral infection.

*Keywords*: avian influenza; antibody; surveillance; environment.

## 1. Introduction

Since the first case of H5N1 was detected in May 1997, as of October 11, 2006 there have been 253 cases of H5N1 worldwide, with the death toll reaching 148 people.[1] The special geographic location of Guangzhou, as well as certain practices of poultry breeding and marketing, and the habits of food handling in Guangdong, have made this region a hot spot of influenza infection. Therefore, since 1997, disease surveillance has been conducted in Guangzhou.

## 2. Results

### 2.1. *Serosurveillance*

Among the total of 2546 samples collected within the period 1997–2005, the positive rate for H9 was 4.56% (116/2546), while the testing for anti-H5 antibody resulted in no positive response. An analysis of 1798 samples showed that the positive rate for anti-H9 antibody was 3.46% for males (43/1241) and 3.95% for females (22/557). No significant difference was found by gender ($\chi^2 = 0.26$; $P = 0.61$). An analysis of anti-H9 positivity is shown in Table 1. The positive rates among different age groups ranged from 2.47% to 4.69%, which was not significantly different ($\chi^2 = 8.43$; $P = 0.13$).

Among serum samples collected in 2006 from 226 workers in the live poultry food markets, where the first H5N1 case was found, one person working in poultry slaughtering was positive for anti-H5 antibody, while another positive was found among the 432 samples from persons without contact with poultry.

The time distribution of anti-H9 positivity was analyzed. The positive rates by year ranged from 0.51% to 10% (Table 2). It is not clear why

**Table 1.** Age distribution of anti-H9 antibody positive rates in surveillance from 2001 to 2005.

| Age group | Number | Positive | Positive rate (%) |
|---|---|---|---|
| 0– | 54 | 0 | 0 |
| 15– | 365 | 13 | 3.56 |
| 25– | 608 | 15 | 2.47 |
| 35– | 493 | 26 | 5.27 |
| 45– | 205 | 8 | 3.90 |
| 55– | 64 | 3 | 4.69 |
| Total | 1789 | 65 | 3.63 |

**Table 2.** Positive rates for anti-H9 antibody in people at high risk of avian influenza from 1997 to 2005.

| Year | *n* | Positive | | Positive rate (%) | |
|---|---|---|---|---|---|
| | | H5 | H9 | H5 | H9 |
| 1997 | 150 | 0 | 15 | 0.00 | 10.00 |
| 1998 | 326 | 0 | 28 | 0.00 | 8.59 |
| 1999 | 99 | 0 | 3 | 0.00 | 3.03 |
| 2000 | 173 | 0 | 5 | 0.00 | 2.89 |
| 2001 | 270 | 0 | 13 | 0.00 | 4.81 |
| 2002 | 207 | 0 | 5 | 0.00 | 2.42 |
| 2003 | 391 | 0 | 2 | 0.00 | 0.51 |
| 2004 | 419 | 0 | 35 | 0.00 | 8.35 |
| 2005 | 511 | 0 | 10 | 0.00 | 1.96 |
| Total | 2546 | 0 | 116 | 0.00 | 4.56 |

there was such a big difference in detecting positive anti-H9 antibodies among different years.

In addition, the positive rates for anti-H9 antibody by occupation were analyzed. The target population was divided into four occupational groups: poultry breeding, poultry selling and slaughtering, animal breeding, and urban people who practiced scattered home-breeding of poultry (Table 3). There was no significant difference in anti-H9 positive rates among the first three groups ($\chi^2 = 3.39$; $P = 0.18$). However, after integrating into a new group, a significant difference was observed between the integrated group (i.e. the first three groups) (anti-H9 positive rate of 4.11%) and the urban group (anti-H9 positive rate of 0.75%) ($\chi^2 = 7.39$; $P = 0.007$).

## 2.2. *Case study*

### 2.2.1. *Case 1: H9N2 case*

A 22-month-old female started to have apparent symptoms on November 5, 1999, received medical treatment beginning on November 6, 1999, and recovered two weeks later.

*Symptoms:* Fever, wheezing and cough.

*Physical examination:* Temperature 38°C; a slight increase in the number of white blood cells.

*Laboratory:* H9 was isolated from her throat swab on the 2nd day by virus culture, and the subtype of H9N2 was identified. Anti-H9N2 antibody was found in the serum sample on the 18th and 31st days after the onset.

*Diagnosis:* Viral tracheitis.

*Close contact:* Anti-H9N2 antibody was found in the mother's serum sample.

*Epidemic statement:* The family, including the parents and the child, were living in a sixth floor apartment in the Baiyun district (a suburb of Guangzhou). They had no live poultry at home and there was no poultry farm nearby. The child sometimes visited a nearby live poultry food market with her mother for shopping, but they had not slaughtered poultry at home. The child had received an influenza vaccination eight months prior to disease onset.

**Table 3.** Anti-H9 antibody positive rates among different groups.

| Year | Poultry breeding | | | Poultry selling & slaughtering | | | Animal breeding | | | Urban people practicing scattered home-breeding | | |
|---|---|---|---|---|---|---|---|---|---|---|---|---|
| | Test number | Positive | Positive rate (%) | Test number | Positive number | Positive rate (%) | Test number | Positive number | Positive rate (%) | Test number | Positive number (%) | Positive rate (%) |
| 2001 | 173 | 3 | 1.73 | 97 | 10 | 10.31 | 0 | — | | 0 | — | — |
| 2002 | 147 | 5 | 3.40 | 60 | 0 | 0.00 | 0 | — | | 0 | — | — |
| 2003 | 173 | 0 | 0.00 | 0 | — | — | 0 | — | — | 218 | 2 | 0.92 |
| 2004 | 176 | 15 | 8.52 | 109 | 14 | 12.84 | 114 | 6 | 5.26 | 20 | 0 | 0.00 |
| 2005 | 113 | 4 | 3.54 | 250 | 4 | 1.60 | 119 | 2 | 1.68 | 29 | 0 | 0.00 |
| Total | 782 | 27 | 3.45 | 516 | 28 | 5.43 | 233 | 8 | 3.43 | 267 | 2 | 0.75 |

#### 2.2.2. *Case 2: First H5N1 case*

A 32-year-old male with disease onset on February 22, 2006, died on March 2, 2006.

*Laboratory testing:* Throat swab and trachea scour detected positive H5 genetic material by means of FQ-PCR, while the serum sample was anti-H5-negative on March 1, 2006.

*Epidemic statement:* This patient was living in an eighth floor apartment in a central district, and had no poultry in the home. The neighbor on the same floor had homing pigeons. The patient did not leave Guangzhou in the recent months before disease onset, but usually visited the live poultry food market, especially during the week before the disease onset. He visited the live poultry food market and was reported as staying for a fairly long time near the same live poultry booth.

#### 2.2.3. *Case 3: H5N1 infection of a poultry slaughterer*

A 44-year-old male presented with sickness on February 26, 2006, and recovered two days later.

*Symptoms:* Cough, nasal congestion, sneezing and fatigue.

*Laboratory testing:* Anti-H5N1 antibody was found in the serum sample on March 5, 2006.

*Epidemic statement:* There was no history of chronic disease, no influenza vaccination, and no contact with animals. The patient was engaged in poultry slaughtering for 4–5 years. During the second year's work at the present booth, he dealt with about 100 poultry per day without protection. None of his workmates showed anti-H5 antibody.

#### 2.2.4. *Case 4: H5N1 infection with no contact with poultry*

A 28-year-old female with antibodies to H5N1 but did not manifest any clinical symptoms.

*Viral laboratory:* Anti-H5N1 antibody was found in the serum sample.

*Epidemic statement*: She was engaged in physical examination in a local district CDC. She had been living in a three-floor villa in the urban community which had a garden with visiting birds. Scattered home-breeding of poultry could be found in the neighbor's home and there was a live poultry food market 50 m away which she sometimes visited. Of the total of five roommates, none presented with any symptoms such as fever and pneumonia. In recent years, she had traveled to other cities, but had not come into contact with sick and/or dead poultry. She had received an influenza vaccination at the end of 2003.

## 3. Discussion

The epidemic of human infections with H5N1 appears to be spreading worldwide. According to a WHO report, there have been 253 persons from 10 countries who have presented with the disease.[1] While the first case in China was reported in October 2005, Guangzhou reported a case in March 2006.[2] Although there is no evidence of effective transmission among human beings,[3,4] research in China indicates that asymptomatic infection may exist among people.[5,6]

H9N2 is one avian influenza virus that humans are susceptible to and has been found in the poultry of Guangzhou.[5] In early 1999, a case of H9N2 infection was reported in a child and the child's mother in Guangzhou. According to the surveillance, the H9 infection rate in the occupational group having direct contact with poultry is 4.56% (0.51%–10.00%), which is higher than that in the urban group that practiced scattered home-breeding of poultry. The variable rate of positive anti-H9 antibodies might be caused by the uneven establishment and adjustment of surveillance stations.

Direct contact with sick and/or dead poultry is regarded as a key risk factor in H5N1 infection. The first case of infection in Guangzhou was not due to direct contact with sick and/or dead poultry; rather, the patient had visited some live poultry food markets.[3] A subsequent swab sample from the live poultry food market concerned tested positive for the H5 virus, and one serum sample from a poultry slaughterer was found positive for anti-H5 antibody.[7] These results indicate that there is some risk of infection in the live poultry market.

There was one H9N2 case and one H5N1 case where the patients did not have any contact with poultry.

In conclusion, exposure to a risk location, such as a food market with live poultry sale and slaughtering or other contaminated places, may be the cause of avian influenza infection in China.

## References

1. World Health Organization, http://www.who.int/csr/disease/avian_influenza/country/cases_table_2006_10_11/en/index.html
2. Taubenberger JK *et al.*, *Nature*, 2005, **437**: 889–893.
3. Wang YL *et al.*, *Chin J Epidemiol*, 2006, **11**: 959–961.
4. World Health Organization, http://www.who.int/csr/disease/avian_influenza/updates/en/index.html:2006-10-09
5. Sun H *et al.*, *Chin J Pub Health*, 2004, **20**(8): 989–990.
6. Li CH *et al.*, *Chin J Exp Clin Virol*, 2004, **18**(13): 213–214.
7. Wang M *et al.*, *Emerg Infect Dis*, 2006, **12**(11): 1773–1775.

# 3.7

# Clinical Significance of Human Cytomegalovirus pp65-Specific Antibodies for Verification of HCMV Primary Infection

Qian Liu, Jing-Xian Chen and Ming-Li Wang*

*Department of Microbiology, Anhui Medical University, The Key Laboratory of Gene Resource Utilization for Genetic Diseases, Ministry of Education and Anhui Province, Hefei 230032, China*

**Objective.** To establish a standard method for clinical diagnosis of human cytomegalovirus (HCMV) primary infection by measuring HCMV pp65 immunoglobulin M (IgM) and the IgG avidity index (AI).

**Methods.** IgM and IgG antibody avidity assays by using protein-denaturing agents and a modified enzyme-linked immunosorbent assay (ELISA) by using recombinant protein pp65 as antigens were studied in 40 children for distinguishing primary HCMV infection from nonprimary infection. Urine samples were inoculated onto human embryo fibroblasts (HFs) after routine treatment for virus isolation. HCMV-specific cytopathic effect (CPE) was observed daily. HCMV UL83 DNA in the cultures was monitored by PCR. HCMV antigen on the slice of infected cells was detected by indirect immunofluorescence.

*Corresponding author. Email: anhuiweishengwu@yahoo.com.cn

**Results.** Among the total of 40 samples, 4 cases which were positive for IgM only were classified as primary infection, 9 cases which were positive for both IgG and IgM were classified as Group A, and the remaining 21 which were positive for IgG only were classified as Group B. Thirty samples classified as HCMV pp65 IgG–positive were measured for AI of HCMV pp65 IgG. AI of HCMV IgG less than 50% was defined as primary infection, 50%–60% was defined as being suggestive of a primary infection, and more than 60% as nonprimary infection. In Group B, there were only 3 cases (14.29%) with AI less than 50%, and the others were all above 60%. While in Group A 5 cases (55.56%) were considered as primary infection, 1 case with AI between 50% and 60% which was suggestive of primary infection and the other 3 cases (62.50%) were classified as nonprimary infection. There were significant differences between Group A and Group B ($P < 0.05$). The concordance rate between the results of the new ELISA and virus isolation was 75.49%. The sensitivity of the new ELISA assay was 100%, the specificity was 87.10% and the accuracy rate was 90%.

**Conclusion.** Detection of HCMV pp65 IgM and HCMV pp65 IgG AI by ELISA is useful for distinguishing primary infection from nonprimary infection, which can be applied for clinical diagnosis.

*Keywords*: human cytomegalovirus; pp65 recombination antigen; antibody; avidity index; primary infection.

## 1. Introduction

Human cytomegalovirus (HCMV) is a ubiquitous herpes virus that is extremely common during perinatal infection. Around 5% of the patients with congenital infection display a series of severe clinical symptoms. The probability of HCMV congenital infection is 0.15%–2.0% in developed countries. It is widely accepted that the risk for fetal infection is greater with maternal primary HCMV infection than with recurrent infection.[1] Therefore, it is of importance to identify HCMV primary infection. Serological diagnosis of HCMV primary infection is based on the detection of specific immunoglobulin M (IgM) antibody. However, recent findings showed that the presence of IgM antibody, which is not always correlated with primary infection, is not sensitive enough for the diagnosis of HCMV infection, and that measurement of the IgG avidity index (AI) is also required. The specificity and sensitivity can be markedly improved by adding the measurement of IgG AI, which is able to identify primary infection.

In this study, HCMV phosphoprotein 65 (pp65)–specific antibody and HCMV pp65 IgG AI measurements were performed in 40 children by an enzyme-linked immunosorbent assay (ELISA) using recombinant protein pp65 as antigens. The results on the single serum samples were compared with those of virus isolation, which served as the gold standard for judging the diagnostic value of the ELISA method.

## 2. Results

### 2.1. *HCMV pp65 antibody measured by indirect ELISA*

Of the 40 samples, 13 were positive for IgM and 30 for IgG. Among them, 4 were IgM-positive only, 9 were positive for both IgG and IgM and were classified as Group A, while the remaining 21 who were positive for IgG only were classified as Group B. Six patients were negative for both IgG and IgM. The cutoff values of AI were established as previously reported by Prince *et al.*[4]: AI < 50% was defined as primary infection, 50%–60% as suggestive primary infection and > 60% as nonprimary infection (including HCMV-reactivated infection and reinfection). As summarized in Table 1, among the 30 cases with positive IgG, 8 had low AI (< 50%), 1 had low AI (50%–60%) and the other 21 had high AI (> 60%); 5 cases in Group B had low AI.

### 2.2. *HCMV virus isolation*

Typical CPE, including cellular swelling, strengthening refraction, and an augmented nucleus which formed megakaryocytic, was observed in HF cells in urine samples among nine cases (Fig. 1). A specific band at 522 bp was observed in PCR detection of HCMV UL83 in cells showing CPE

**Table 1.** Results of the HCMV pp65 IgG avidity index by ELISA.

| Group | $n$ | AI | | |
|---|---|---|---|---|
| | | < 50% | 50%–60% | > 60% |
| | | Cases (%) | Cases (%) | Cases (%) |
| Positive IgG and IgM (Group A) | 9 | 5 (55.56) | 1 (11.11) | 3 (33.33) |
| IgG-positive only (Group B) | 21 | 3 (14.29) | 0 (0) | 18 (85.71) |
| IgM-positive only | 4 | ND* | ND | ND |

*ND: no detection. There was a significant difference between Group A and Group B by $\chi^2$ analysis ($\chi^2 = 8.239$; $P < 0.05$).

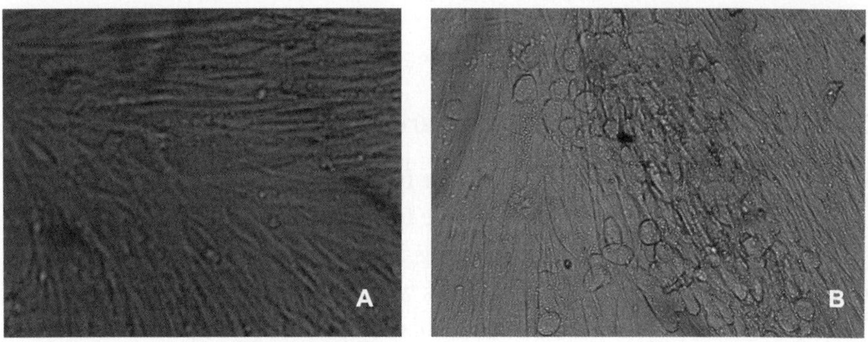


**Fig. 1.** Virus isolation of the urine samples (×200). (A) Normal HF cells. (B) Cytopathic effect on HF cells inoculated in the urine samples.

(Fig. 2). The positive green signal was observed in HF cells infected with HCMV by indirect immunofluorescence, while the results of other cases were all negative.

## 2.3. *Calculation of the kappa value*

The results of virus isolation were taken as the gold standard, and results of the two methods are compared in Table 2.

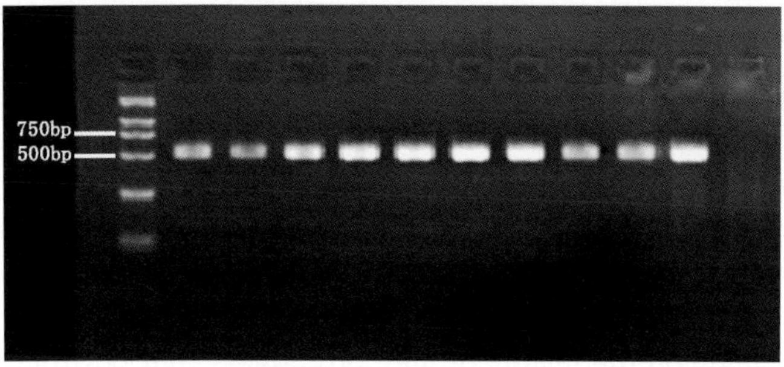


**Fig. 2.** Detection of UL83 in CPE-positive samples with PCR. M — DNA marker; 1 — positive control; 2–10 — positive samples; 11 — negative control.

**Table 2.** Comparison of the results between HCMV pp65 IgM and virus isolation.

| pp65 IgM | Virus isolation | | Total |
|---|---|---|---|
| | + | − | |
| + | 9 (a) | 4 (b) | 13 ($\gamma_1$) |
| − | 0 (c) | 27 (d) | 27 ($\gamma_2$) |
| Total | 9 ($C_1$) | 31 ($C_2$) | 40 (N) |

Kappa = 75.49%; sensitivity = 100%; specificity = 87.10%; accuracy = 90.00%.

## 2.4. *Judgment of primary infection*

Of the total of 40 samples, 4 that were positive for IgM only with virus isolation were classified as HCMV primary infection. Of the 30 children who were IgG-positive, 8 with a low AI (< 50%) were defined as HCMV primary infection. Among these 8 children, 5 who had positive IgG and IgM with virus isolation were defined as HCMV primary active infection, while the other 3 who were negative for IgM without virus isolation were considered as HCMV latent infection, who could have been infected with HCMV before. One case with AI between 50% and 60% was considered as suspicious primary infection.

## 3. Discussion

The incidence of congenital infections with HCMV primary infection has increased dramatically and congenital infections of HCMV can cause damage in multiple systems. Therefore, distinguishing HCMV primary from nonprimary infection in the early clinical stages of pregnancy is very important. However, currently, there is no reliable diagnostic standard for HCMV primary infection. The detection of HCMV primary infection mainly depends on the results of antibody assays, but positive IgM antibody responses cannot represent genuine HCMV primary infection, due to cross-reaction. Therefore, we have to combine other indexes to supplement and identify primary infection. One of the most effective methods is to combine the detection of HCMV IgG-AI[2] with ELISA. The traditional ELISA displays low specificity using nonprocessed HCMV antigen, which cannot reflect the status of HCMV infection accurately. Virus isolation, which can be employed as a gold standard for HCMV active infection, is difficult to carry out for clinical use due to its time-consuming nature and the requirement for a cell culture facility.

HCMV pp65 is a structural protein expressed in the early stage of HCMV infection. It is expressed in peripheral blood lymphocytes, monocytes, polymorphonuclear leucocytes and vascular endothelial cells 6–24 h after infection. HCMV pp65 is an early index for indicating HMCV active infection. The detection of pp65 antigenaemia is a relative gold standard for diagnosis of active infection. In addition, pp65 is not only the main antigen in antigenemia, but also the major target for HCMV-specific cytotoxicity T cells (CTLs). Pp65 can induce both specific antibodies and cell-mediated immune response. It has also been reported that this protein is the main recognition protein and effective antigen for Th and T memory cells.[3] Therefore, pp65 has been the candidate for vaccine development and immunotherapy. HCMV pp65 can induce strong and specific IgM response in primary infection; however, since IgM is taken as a marker of recent infection, indirect ELISA was used in this study. The homemade truncated pp65 recombinant protein was utilized for detecting pp65-specific antibody and IgG AI. Results of this antibody assay were compared with traditional virus isolation in order to determine its feasibility in detecting infection.

Detection of IgG-AI has been reported as a good method for distinguishing primary infection from nonprimary infection.[4] This method is easy to perform; it needs only one additional step of degradation in the conventional indirect ELISA for detecting IgG and it is effective for identifying the status of infection by combining IgG AI and IgM.[2] As for pregnant women, the accurate diagnosis of HCMV primary infection should be evaluated globally, including serological detection, pp65 antigenemia, virus detection and clinical manifestations.[5] Such studies should be conducted with extraordinary care and concern.

In this study ($n$ = 40 cases), combining the two assay methods and using virus isolation as the gold standard, it was observed that 5 cases (55.56%) had an AI less than 50% and were classified as true primary infection, while 1 case had an AI between 50% and 60%, and was judged as suspicious primary infection, because of negative virus isolation. The other 3 cases (62.50%) with AI above 60% were nonprimary infection. Although there were only 3 cases (14.29%) in Group B with AI less than 50%, the virus isolation from these cases was negative and they were considered to be HCMV primary infection before long, but the specific IgM had converted to being negative and the virus was in latency.

It has been documented that the primary infection rate of pregnant women is 0.89%, while the noninfection rate is 11.11%. According to the reports published in the international literature,[6] the determination of HCMV IgG AI at 6–18 weeks' gestation could identify all women who would have an infected fetus/newborn (100% sensitivity), whereas IgM detection yielded poorer results (69% sensitivity). Interestingly, at 20–23 weeks' gestation, the sensitivity of IgM detection was higher than that obtained by AI (75% and 63%, respectively) and the combination of IgG AI and IgM yielded the best results (81% sensitivity). We used homemade pp65 recombinant protein for the detection of HCMV pp65-specific antibody and IgG AI. HCMV pp65 recombinant protein could take the place of HCMV for serological detection, and the combination of IgG AI and specific antibody could be one of the effective ways for the diagnosis of HCMV primary infection. In our study with a limited number of patients, we showed that this method has high sensitivity and specificity, and the concordance rates can reach 75.49% compared with traditional virus isolation. Due to its convenience and because there is no requirement for special equipment, we expect that this method can be used in hospitals in China with varying levels of technology.

## References

1. Ergun UG *et al., Saudi Med J*, 2007, **28**(2): 264–267.
2. Baccard-Longere M *et al., Clin Diagn Lab Immunol*, 2001, **8**(2): 429–431.
3. Ghanekar SA *et al., Clin Diagn Lab Immuol*, 2001, **8**(3): 628–631.
4. Prince HE *et al., Clin Diagn Lab Immunol*, 2002, **9**(4): 824–827.
5. Revello MG *et al., Clin Microbiol Rev*, 2002, **15**(4): 680–715.
6. Lazzarotto T *et al., Viral Immunol*, 2000, **13**: 137–141.

# 3.8

# Analysis of Indeterminate HIV Western Blot Profiles

Kai Gao, Yan Li*, Cai-Yun Liang, Hui-Fang Xu and Zhi-Gang Han

*Guangzhou Center for Disease Control and Prevention, Guangzhou 510080, China*

**Objective.** To study the characteristics of and reasons for the indeterminate HIV Western blot (WB), the limitation of WB and the possible amendatory measures.

**Methods.** Summing up and analyzing the distribution of the indeterminate HIV WB among the tested people, the laboratory test results, the follow-up and the final HIV antibody outcome.

**Results.** The relatively healthy people, including the people involved in voluntary counseling and testing, the blood donors and the pregnant women, account for 50% of the indeterminate HIV WB. The follow-up for the indeterminate HIV WB is difficult and few people with the indeterminate HIV WB have an HIV antibody test again. There is a false positive in the WB test; in particular, the false positive of P24 is serious.

*Corresponding author: Email: liyanwb@163.com

**Conclusion.** The indeterminate HIV WB is related to the false positive of the WB test. Measures should be taken to reduce the indeterminate HIV WB and the results should be interpreted accurately.

*Keywords*: human immunodeficiency virus; Western blot; false positive; indeterminacy; follow-up.

## 1. Introduction

To date, the standard method of detecting human immunodeficiency virus (HIV) infection to test HIV antibody, including two steps, namely the primary screening test and the confirmation test. There are several primary screening methods, but the most commonly used confirmation test is Western blot (WB). The basic principle of WB is to separate the whole HIV virus antigen after SDS–acrylamide gel electrophoresis (SDS–PAGE) according to relative molecule weight and then transfer it to cellulose nitrate membrane. Diluted serum samples will then be added onto the cellulose nitrate membrane. If the serum contains anti-HIV antibodies, these antibodies will specifically bind to their ligand, causing the appearance of purple–brown bands after enzyme-linked reaction, for the inspection of results. WB data will bring out three possible results: "HIV-negative," "HIV-positive" and "HIV-antibody-indeterminate." In the management of the epidemic, "HIV-positive" is relatively simple and clear while "HIV-antibody-indeterminate" is required to conduct follow-up so as to determine the final outcome.[1] In recent years, many reports and our daily tests[2] have indicated the increased rate of "HIV antibody indeterminacy," which has not only brought many problems to laboratory investigation and epidemic management of HIV/AIDS, but also imposed a heavy mental burden on individuals. To analyze the reasons, the distribution characteristics in the population, the specificity of laboratory test results of HIV indeterminate, follow-up and rechecking were carried out. This study, based on our data on HIV antibody indeterminacy and the relative information about the samples examined during 2004–2005, will analyze the characteristics and the reasons for the indeterminacy and provide possible measures for reducing it.

## 2. Results

### 2.1. *General profiles and follow-up information on cases with HIV antibody indeterminacy*

In May 2004 our laboratory was approved as one of the institutes responsible for HIV antibody confirmation tests in Guangzhou. There were a total of 44 cases with HIV antibody indeterminacy, including 13 cases (1.34%) in 2004 and 31 cases (2.4%) in 2005, with a significantly increased number in 2005 compared to that for the same period in 2004. Among 44 cases of HIV antibody indeterminacy, only 10 cases (22.7%) had a clear risk behavior and the remaining 34 cases (77.3%) were unknown (Table 1). These cases included 26 men and 19 women, with volunteers and blood donors constituting a relatively large proportion. In accordance with the "National AIDS Testing Technical Specifications, 2004," a person with HIV antibody indeterminacy should be followed up every three months, a total of two times. In this study, there were only 14 cases (31.8%) successfully followed up and rechecked. Among the various groups, the volunteers had the highest rate of being reexamined. The distribution is presented in Table 2.

**Table 1.** The possible transmission distribution of the indeterminate HIV WB.

| Possible transmission | Sexual contact | IDU | Blood/ blood product | Vocation exposure | MICT | UK | Total |
|---|---|---|---|---|---|---|---|
| *n* | 5 | 1 | 2 | 1 | 1 | 34 | 44 |
| (%) | (11.36) | (2.27) | (4.54) | (2.27) | (2.27) | (77.3) | (100) |

IDU — injection drug use; MICT — mother-to-child transmission; UK — unknown.

**Table 2.** The people distribution of the indeterminate HIV WB and the callback.

| People | VCT | Blood donor | Pregnant/ lying-in woman | Others | Total |
|---|---|---|---|---|---|
| *n* (component, %) | 11 (25) | 9 (20.45) | 2 (4.65) | 22 (50) | 44 (100) |
| *n* (callback, %) | 9 (81.8) | 1 (11.1) | 1 (50*) | 3 (13.6) | 14 (31.8) |

*It is unable to analyze because the sample volume is too small.

VCT — voluntary counseling and testing.

**Table 3.** The WB band distribution of the indeterminate HIV WB.

| WB brand | *n* (%) | Note |
|---|---|---|
| P24 | 23 (52.27) | Only 20 samples have P24 |
| gp160 | 17 (38.63) | Only 13 samples have gp160 |
| gp120 | 3 (6.81) | |
| P66 | 3 (6.81) | |
| P17 | 3 (6.81) | |
| P51 | 2 (4.54) | |
| P31 | 1 (2.27) | |

## 2.2. *Data on HIV antibody indeterminacy*

There were a total of 44 HIV-antibody-indeterminacy samples, including 17 samples with PA negative but ELISA positive, and 8 samples with PA positive and ELISA positive. The cases of HIV-antibody-indeterminacy WB-confirmed bands were mainly concentrated in the P24 and gp160 bands; the results on specific distribution are shown in Table 3.

## 2.3. *Follow-up results on the cases with HIV antibody indeterminacy*

In accordance with the "National AIDS Testing Technical Specifications, 2004," the subjects with HIV antibody indeterminacy should be followed up every three months, two times. If the results are still suspicious or negative, negative reports will be given, and vice versa. Only 14 cases had been followed up. One case was confirmed positive; 4 cases were still indeterminate for the first time, but they did not continue with the second test, which brought us difficulty in making an accurate diagnosis. The other 9 cases were confirmed negative, including 8 cases with P24 positive (Table 4).

# 3. Discussion

In accordance with the "National AIDS Testing Technical Specifications, 2004," individuals with HIV antibody indeterminacy should be

**Table 4.** The follow-up callback test results of the indeterminate HIV WB.

| *n* | Original WB band | Follow-up callback test results (WB band) |
|---|---|---|
| 1 | gp120 | Positive (gp160 gp120 gp41 p66 p55 p51 p39 p31 p24 p17) |
| 3 | gp160 | Indeterminate (gp160) |
| 1 | P24 | Indeterminate (P24) |
| 1 | gp160, gp120 | Negative |
| 6 | P24 | Negative |
| 2 | P24, P17 | Negative |

followed up. In this study, only 14 of the 44 cases (31.8%) were followed up, which caused a low overall follow-up rate and poor compliance. For those without follow-up, the final outcome could not be determined, the epidemiological investigation and standardized management could not be done, thus increasing the risk of hidden spread of AIDS. However, as mentioned worldwide, among all kinds of people, those who voluntarily asked for counseling and tests had a higher rate of follow-up, which indicates that counseling about AIDS could increase the rate of follow-up and compliance. With the increased intensity of monitoring AIDS, some of the general healthy populations were also included in the monitoring of HIV infection, such as patients in hospitals and pregnant women, in addition to the high-risk groups such as intravenous drug users, females in the entertainment business and gay people. In 14 follow-up cases, 1 case with HIV antibody indeterminacy in a preoperative examination was confirmed HIV-antibody-positive after follow-up, which shows the importance of broadening the spectrum of HIV screenings to decrease the AIDS epidemic and to find early AIDS patients. In this study, people voluntarily asking for counseling and tests, blood donors and pregnant women accounted for 50% of the subjects with HIV antibody indeterminacy. Given that these three categories of people belonged to the healthy population, and in the light of the relatively low level of HIV infections among the general population in China, a number of the confirmation results could be false positive.

In fact, despite the good specificity of WB, false positive results often occurred, and conclusions should be drawn carefully. In a survey of 13 cases without high-risk factors, WB results showed positive reactions versus viral protein bands and the sera were considered as HIV-1-positive but were actually negative by further tests.[3] In 1993, in a retrospective study on people in the low-risk groups, the sera reacted with the envelope protein "gp41 + gp160/gp120" or gag-coated (P24 + gp41 and or gp160/gp120) zones, and were judged as HIV-1-positive but were in fact negative by follow-up.[4,5] In addition, some data showed that the incidence of reaction with P24 in the serum of HIV-1 infection was 38% higher than that of other viral antigens, and thus is crucial for early detection of HIV.[6] On the other hand, the high sensitivity of reaction with P24 could also give the WB confirmation test the risk of a false positive result. Furthermore, a study reported that among 12 500 samples, 15 showed an abnormal band by WB, and were grouped as HIV-antibody-indeterminate. In the aberrant zones, more than half of them were P24-based, and most of them with only P24, but P24 antigen tests and PCR tests were negative in these samples.[7] Our study is similar to previous studies in the WB band distribution. P24 and gp160 constituted the main type of indeterminacy in the HIV antibody zone, in which antibodies reacting only with P24 accounted for 45.5%. In the group with HIV antibody indeterminacy, 50% were in the relatively healthy population. Besides, P24 was the main band in the false positive samples, based on follow-up data, which confirmed the WB false positive by the appearance of the nonspecific P24 band.

The phenomenon of HIV antibody indeterminacy occurring in relatively healthy groups (the people voluntarily asking for counseling and tests, blood donors and pregnant women) and some patients in hospital has challenged the accuracy of WB data. The reasons could be as follows:

(1) The presence of autoantibody.
(2) A false positive by vaccination. A study found that 1–2 years after vaccination versus influenza nonspecific positive responses to many viruses could be induced, including HIV-1. In addition, vaccination

with the HIV-1 subunit vaccine also could render the WB test positive, but this should not lead to the conclusion of HIV-1 infection.

(3) The false positive caused by some diseases, such as collagen-vascular diseases, autoimmune diseases, cancer, malaria, kidney disease, tuberculosis, viral infection and common antibody cross-reaction could make false positives occur in the WB test.[8–11]

(4) A false positive caused by special physiological conditions, such as pregnancy and allergies. Previous study showed that serum from HIV-negative healthy pregnant women reacted positively with the gp41 band.[12] Our lab also found a P24-band-positive pregnant woman with HIV antibody indeterminacy, but her antibody reaction turned negative in the follow-up test. All in all, in the course of HIV detection, in addition to reports from laboratory tests, one should also do a comprehensive analysis of the history of epidemiology.

The result of HIV antibody indeterminacy makes AIDS epidemiological investigation and epidemic management rather complicated, and imposes a heavy psychological burden on the examined individual. Therefore, laboratories should take appropriate measures to minimize the possibility of HIV antibody indeterminacy and provide accurate interpretation of the results. These would include: (1) being alert to possible false positive reactions with the P24 band; (2) employing assisted examinations, such as HIV-1 RNA quantitative detection, the direct detection of HIV-1 nucleic acid rather than antibodies; using the flow cytometry method for detecting HIV-1 bands (gp160, gp120, gp41 and P24) occurring in WB; accurate quantitation and localization; HIV reverse-transcriptase activity assay; and (3) repeat WB tests.

## References

1. China Ministry of Public Health, "National AIDS Testing Technical Specifications," 2004.
2. Liu QZ *et al.*, *J Sin Acta Dermatol*, 2004, **37**: 245–246.
3. Cordes RJ *et al.*, *Postgrad Med*, 1995, **98**: 177–180.
4. Healey DS *et al.*, *AIDS*, 1993, **7**: 655–658.

5. Sayre KR *et al.*, *Transfusion*, 1996, **36**: 45–52.
6. Busch MP *et al.*, *Transfusion*, 1996, **36**: 41.
7. Migali E *et al.*, *Allergol Immunopathol* (*Madr*), 1993, **21**: 61–65.
8. Walensky RP *et al.*, *Clin Infect Dis*, 2001, **33**: 570–572.
9. Ghosh K *et al.*, *Br J Biomed Sci*, 2001, **58**: 20–23.
10. Silverstein DM *et al.*, *Pediatr Nephrol*, 2004, **19**: 547–549.
11. Esteva MH *et al.*, *Ann Rheum Dis*, 1992, **51**: 1071–1073.
12. Myrmel H *et al.*, *APMIS*, 1988, **96**: 950–951.

3.9

# Detection of Multidrug-Resistant *Mycobacterium tuberculosis* by Sputum Smears

Shu-Yan Wu*, Guo-Hao Gu† and Rui Huang*,‡

**Department of Microbiology, Medical College of Soochow University, Suzhou 215123, China*

*†Department of Medical Laboratory Science, The First Affiliated Hospital of Soochow University, Suzhou 215006, China*

**Objective.** To explore the possibility of determining multidrug-resistant *Mycobacterium tuberculosis* (MDR-TB) directly from acid-fast bacillus (AFB) stain–positive sputum smears.

**Methods.** A total of 18 isolates of *M. tuberculosis* with known phenotypic rifampin-resistant profiles and the original stained sputum smears were collected from three hospitals. DNA was extracted from both the isolates and the stained sputum smears, and the *rpoB* gene was amplified by PCR. Identification of rifampin-resistance-related mutations in the *rpoB* gene was performed by hybridization using line probe assay and DNA sequencing.

**Results.** Sixty-seven percent (12/18) of the isolates were resistant to rifampin. Among the 12 resistant strains, *rpoB* gene mutations were detected in 12 (100%) DNA samples extracted from isolates and in

‡Corresponding author. Email: hruisdm@163.com

11 (91.7%) from stained sputum smears. Results obtained by line probe assay and sequence analysis for rifampin resistance determination were identical.

**Conclusion.** Detection of rifampin-resistance-related mutations in the *M. tuberculosis rpoB* gene directly from AFB stain-positive sputum slides could provide a simple and rapid tool for clinical laboratory analysis.

*Keywords*: *Mycobacterium tuberculosis*; drug resistance; sputum smears.

## 1. Introduction

Tuberculosis (TB) is a major infectious disease and an important cause of death throughout the world. According to World Health Organization (WHO) statistics, there are 9.2 million new cases each year, and around 1.7 million deaths per year. Ninety-nine percent of the deaths occurred in developing countries.[1] TB morbidity and mortality rates are rising, mainly due to the broad distribution and rapid spread of multidrug-resistant tuberculosis (MDR-TB), which has become a common global challenge. However, currently only 2% of the MDR-TB cases worldwide are diagnosed and treated, largely due to inadequate laboratory services.

In recent years, scientists have focused on the resistance mechanism of *Mycobacterium tuberculosis* using molecular biology techniques. These techniques have promoted rapid identification methods for resistant mutant strains by identifying the location of *M. tuberculosis*-resistant genes and gene mutations. Rifampin (RFP) is a broad spectrum antibiotic which can integrate with the RNA polymerase β subunit, thereby inhibiting the mRNA transcription. Almost all RFP-resistant bacteria are caused by the mutation of a particular region of the *rpoB* gene (RNA polymerase subunit β gene), which leads to the RNA polymerase β subunit no longer binding with the RFP. There is a 69-bp hypermutation region near the center of the *rpoB* gene of RFP-resistant *M. tuberculosis*, and up to 95% of drug-resistant *M. tuberculosis* is due to missense, deletion or insertion mutations.[2] As a high mutation rate has been observed in this hypermutation region and single RFP-resistant strains are extremely rare, detecting *rpoB* mutations has become the symbol of MDR-TB.[3] In this study, in order to detect MDR-TB, the identification of the mutant *rpoB* gene was performed by hybridization using line probe assay and DNA sequencing. The results are as follows.

## 2. Results

### 2.1. M. tuberculosis *drug susceptibility test*

Among the 18 isolates, 67% (12) were resistant to RFP, and the other 33% (6) were sensitive.

### 2.2. *Solid phase hybridization of the PCR-amplified* rpoB *fragment and probes*

In 12 resistant strains, analysis of the *rpoB* gene mutant revealed that DNA samples from the cultures were 100% (12/12), whereas 1 case (R7) in their corresponding stained sputum slides could not be detected. The positive rate of *rpoB* mutant detection was 91.7% (11/12). The hybridization of the PCR-amplified *rpoB* fragment and probes showed that 4 mutants (531 TTG, 531 TGG, 526 TAC and 516 TTC) could be detected in wild-type-to-mutant (W → M) changes. However, R2 and R9 mutations (516 GTC) were only found in negative hybridization with W2, and it was not clear what mutations were responsible for the M type (Table 1). Among 6 sensitive strains, no *rpoB* mutants were detected.

### 2.3. rpoB *DNA sequencing*

Nested PCR amplification was used in our study, because amplified DNA from the sputum smear was less than enough. Sequence analysis of the *rpoB* mutants in cultured bacterial cells and their corresponding stained sputum slides were identical, presenting the samples containing RFP-resistant *M. tuberculosis* carried altered codons within the core region of the *rpoB* gene, which demonstrated mutation at the 516 site (GAC → GTC, Asp → Val, and GAC → TTC, Asp → Phe) in four strains, mutation

**Table 1.** *rpoB* PCR fragment hybridization with probes and DNA sequence analysis.

| Strain | Source | Southern blot | | *rpoB* mutant | |
|---|---|---|---|---|---|
| | | Strain | Slide | Strain | Slide |
| R1 | Canada | W5–/M1+ | W5–/M1+ | 531-TTG | 531-TTG |
| R2 | Canada | W2–/– | W2–/– | 516-GTC | 516-GTC |
| R3 | Canada | W5–/M2+ | W5–/M2+ | 531-TGG | 531-TGG |
| R4 | Anhui | W2–/M1+ | W2–/M1+ | 516-TTC | 516-TTC |
| R5 | Anhui | W5–/M1+ | W5–/M1+ | 531-TTG | 531-TTG |
| R6 | Anhui | W5–/M1+ | W5–/M1+ | 531-TTG | 531-TTG |
| R7 | Anhui | W2–/M1+ | –/– | 516-TTC | –/– |
| R8 | Anhui | W4–/M4+ | W4–/M4+ | 526-TAC | 526-TAC |
| R9 | Jiangxi | W2–/– | W2–/– | 516-GTC | 516-GTC |
| R10 | Jiangxi | W5–/M1+ | W5–/M1+ | 531-TTG | 531-TTG |
| R11 | Jiangxi | W5–/M2+ | W5–/M2+ | 531-TGG | 531-TGG |
| R12 | Jiangxi | W5–/M1+ | W5–/M1+ | 531-TTG | 531-TTG |
| S13–18 | Three places | W1–5+<br>M1–4– | W1–5+<br>M1–4– | Wt | Wt |

R — RFP-resistant strain; S — RFP-sensitive strain. W1–5 — *rpoB* gene wild type probes; M1–4 — *rpoB* gene mutant probes. –, +: The results of hybridization with the corresponding probes.

at the 526 site (CAC → TAC, His → Tyr) in one strain, and mutation at the 531 site in seven strains (TCG → TTG, Ser → Leu, and TCG → TGG, Ser → Trp) (Table 1).

## 3. Discussion

An important aspect of studying the molecular mechanism of *M. tuberculosis* drug resistance is not only the establishment for rapid identification of *M. tuberculosis*-resistant gene types, but also knowledge about the drug resistance of isolates from TB patients. This is the guidance on effective clinical chemotherapy and on the prevention of further spread of resistant bacteria. Analysis of drug resistance genes does not require the culture of *M. tuberculosis*, and therefore it has many advantages over the traditional *M. tuberculosis* drug susceptibility test. The advantages include a decrease in the detection time from a few months or weeks to a few days, and lowering the risk of contamination by partially automated operation aids in the laboratory. In this study, amplifying *rpoB* gene fragments by PCR, hybridization with a series of probes, we employed DNA sequencing at length and comparison of the resistance gene of *M. tuberculosis* from clinical acid-fast bacillus stain–positive sputum smears with cultured bacteria, and we explored the feasibility of the former method for direct use in the clinical laboratory.

The results demonstrated that positive acid-fast bacillus stains of sputum smears from clinical samples can be directly used for hybridization with probes after PCR amplification, to test whether there are drug-resistant strains. The detection rate can reach 91.78% (11/12). In 12 samples, 4 kinds of mutants (516 TTC, 531 TTG, 531 TGG and 526 TAC) could be identified with probes. However, R2 and R9 mutations (516 GTC) could only be found in negative hybridization with W2, and it was not clear what mutations were responsible for the M type. MDR-TB was found as long as the *rpoB* gene mutation could be detected for actual application. *rpoB* mutant DNA sequencing results of 12 cultured strains and 11 sputum smears showed that PCR amplification of DNA from both strain and slide hybridization with line probes was identical. All drug-resistant strains had changed nucleic acid sequences: serine 531 → leucine, tryptophan (Ser531 → Leu, Ser531 → Trp) and histidine 526 → tyrosine (His526 → Tyr). Two mutations accounted for about 67% of the total number of mutations (8/12), similar to other laboratory reports.[4] The results of this study demonstrated that positive acid-fast bacillus stain samples of sputum smears could be directly used for detection of the *M. tuberculosis* RFP-resistant gene. This approach avoided laboratory infections when using viable bacteria, and could be completed within one day. It was time-saving, economic, safe and accurate, and a suitable method for the clinical laboratory.

## References

1. WHO Report 2008, Global tuberculosis control–surveillance, planning, financing (WHO/HTM/TB/2008.393).
2. Bartai Z *et al.*, *J Clin Microbiol*, 2001, **39**: 3736–3739.
3. van der Zanden AGM *et al.*, *J Clin Microbiol*, 2003, **41**: 1101–1108.
4. Patnaik M *et al.*, *J Clin Microbiol*, 2001, **39**: 51–52.

SECTION 4

# Clinical Medicine

# Introduction and Comments: A Glimpse of Clinical Infectious Diseases in China in the New Millennium

Shan Lu

*Department of Medicine, University of Massachusetts Medical School, 55 Lake Avenue North, Worcester, MA 01655, USA*

While China has seen a significant increase in cardiovascular diseases and malignancies as the major causes of mortality in the last several decades, infectious diseases remain a prime threat to the people's health, which may be expected from a country that claims to be the largest developing country in the world. A new generation of infectious disease specialists has been trained with a focus on the diagnosis and treatment of hepatitis B virus infection, which is considered the foremost infectious disease in the nation. This training process has created a knowledge gap in many infectious disease specialists who are less experienced in identifying and managing less common infectious diseases that are still present but at a lower level. Social and economic changes have led to a new environment that will no doubt constantly reshape the spectrum of infectious diseases in a population of 1.3 billion.

Thus, it is quite remarkable that the founders of the *Journal of Microbes and Infection* (*JMI*) had a vision of putting significant emphasis on submitted papers related to clinical research. While there are other dedicated clinical journals in China, including a few specializing in infectious diseases, the *JMI* publishes clinical studies related to less popular topics in an attempt to identify any new research that may indicate a new trend in infectious diseases. One example of such research is the clinical analysis of 77 measles patients by Shi and Xie (Chapter 4.3). While this study only summarized measles cases in one hospital in Shanghai, it is still quite shocking to see that there were 77 cases of this disease in just one year (2005) in a group mainly consisting of adults (94.8% being 20–49 years old). Data revealed that most of the patients in this group were non-Shanghai residents, indicating that they were most likely migrant workers from less-developed regions where required measles immunization campaigns may have much less coverage than in Shanghai. This report not only teaches less experienced infectious disease physicians to think about the possibility of measles in patients with fever and abnormal liver function tests, but also provides a strong argument for government agencies to establish an improved vaccination requirement for adults who may not have a documented history of measles vaccination.

Two reports focus on infections in children. One (Shen *et al.*, Chapter 4.5) looks at the etiology and clinical characteristics of children hospitalized for acute lower respiratory tract infection during winter. RSV continues to be the main cause of such infection, followed by influenza type A and type B, parainfluenza viruses and adenovirus. This information is reassuring, in that no emerging infections are becoming prominent in children. The other report (Teng *et al.*, Chapter 4.2) identifies the genotype of norovirus detected in diarrhea specimens from children in a hospital in Shanghai. Detailed sequence analysis allowed the separation of different samples into different subgroups. Given the fact that diarrhea remains a major disease in China, this study was very informative in providing a more complete picture of the causes of diarrhea in different age groups in China.

The report by Zhang *et al.* (Chapter 4.1) on a very common disease, chronic obstructive pulmonary disease (COPD), seeks to determine the roles of lower airway bacterial colonization in the exacerbation of COPD. The literature on the use of antibiotics for the control of severe COPD is inconsistent, since the connection between bacterial infections and COPD is unclear. The present study identified key colonized bacteria in COPD, and selected cytokines were measured in patients' sputum. While future studies are clearly needed to provide a more in-depth analysis, this report makes progress in our understanding of the factors that may affect the course of COPD.

Moving to the recent earthquake in Sichuan, a paper (Chapter 4.6) describes infectious diseases and wound infections. Experiences in treatment of wound infections in four cases with multiple injuries are presented, and one special case is reported in detail. As most clinicians concentrate more or less on their own specialties, the importance of collaboration between surgeons, physicians and microbiologists is emphasized.

China, like many other countries in the world, cannot escape HIV-1 infection, the worst pandemic of the 20th century. Gong *et al.* (Chapter 4.4) describe the effectiveness of screening and confirmation testing for detection of HIV infection in China. A key finding involves delayed detection of HIV infection, which is very possible if a clinician does not consider HIV-1 while purely treating various opportunistic infections. This report is effective in confronting the same HIV issues in China as reported in other countries, which may be important for health policy decision-makers to develop a public health strategy that can achieve the best control of HIV-1 infection in China.

Many people in the world will immediately think of SARS or avian flu when the topic of China's infectious diseases is raised. The reality is that China is similar to many other countries, in that it has its own unique set of infectious disease issues, and the spectrum of infectious diseases changes as society progresses. In the 21st century, emerging or re-emerging infectious diseases remain as key concerns, but understanding the baseline spectrum of infectious diseases in China will provide critical knowledge of how routine and new infectious diseases are identified and treated. Selected clinical papers from the *JMI* in this book provide a unique opportunity for the world to take a closer look at infectious disease practices in China.

# 4.1

# The Role of Lower Airway Bacterial Colonization at the Stable Stage of Chronic Obstructive Pulmonary Disease

Min Zhang, Xin Zhou*, Xin-Yi Zhang and Xing Ding

*Department of Pulmonary Medicine,*
*Shanghai First People's Hospital, Jiao Tong University,*
*Shanghai 200080, China*

**Objective.** To test the hypothesis that lower airway bacterial colonization (LABC) existed in the stable phase of chronic obstructive pulmonary disease (COPD) and to correlate LABC with airway inflammation and exacerbation frequency.

**Methods.** A total of 46 patients with moderate-to-severe COPD were included in this study. Their sputum and bronchoalveolar lavage (BAL) specimens were collected at the first visit as the baseline and at the second visit at the exacerbation period. The samples were analyzed for bacterial growth by culture, and the levels of IL-6, IL-8 and TNF-α were measured by ELISA.

**Results.** Among 46 subjects, 15 (32.6%) and 21 (45.7%) had LABC with bacterial loads >$10^6$ CFU/ml in their sputum specimens at visits

*Corresponding author. Email: xzhou53@163.com

1 and 2, respectively. Bacterial counts $>10^{3.0}$ CFU/ml in BAL specimens were recovered in 44.4% (4/9) and 54.5% (6/11) at visits 1 and 2, respectively. *Haemophilus influenzae* was the predominant pathogenic organism isolated. IL-6 and TNF-α in the sputum of patients with LABC were significantly higher than in those without LABC. The presence of LABC had a significant positive correlation with elevated IL-8 at visit 1 ($P < 0.05$).

**Conclusion.** LABC was present in certain subpopulations of stable COPD patients. It may be responsible for COPD exacerbation by increasing bacterial colonization and/or promoting airway inflammation.

*Keywords*: chronic obstructive pulmonary disease; bacterial colonization; cytokine.

## 1. Introduction

Chronic obstructive pulmonary disease (COPD) is a disease characterized by airflow limitation that is not fully reversible. The airflow limitation is usually both progressive and associated with an abnormal inflammatory response of the lung to noxious particles or gases.[1] There are two phases to COPD patients: stable stage and acute exacerbation. COPD patients are usually transformed from the stable stage into acute exacerbation. Acute exacerbation is defined as the worsening of at least two of the following respiratory symptoms: cough, sputum production, and wheeze, dyspnea on exertion and chest tightness, which beyond normal day-to-day variations is acute onset and may warrant additional treatment.[2] Exacerbation frequency is now recognized as an important feature of the natural history of COPD, and also has important implications for health-related quality of life.[3] The pathogenesis of acute exacerbation of COPD (AECOPD) is complicated. It was recently demonstrated that lower airway bacterial colonization (LABC) was often present in COPD patients at the stable stage,[4] but the relationship between LABC and AECOPD was still largely unknown. In this study, we aimed to evaluate the relationship between LABC and airway inflammation or exacerbation frequency by means of quantitative identification of bacteria and measurement of the levels of IL-6, IL-8 and TNF-α in sputum and bronchoalveolar lavage (BAL) specimens at the stable and the exacerbation stages; the knowledge of this relationship may contribute to further understanding of the pathogenesis of AECOPD.

## 2. Results

### 2.1. *Patients' characteristics*

The baseline physiologic characteristics of the 46 patients who were recruited for the study comprising 38 males and 8 females, are summarized in Table 1. The patients were classified as suffering from moderate-to-severe COPD according to the Chinese Initiative for Chronic Obstructive Lung Disease criteria; they were made up of 24 moderate patients (52.1%) and 22 severe patients (47.9%). The mean age was 72.02 ± 5.64, the mean value of the body mass index (BMI) was 22.74 ± 4.28 kg/m$^2$, the mean value of $FEV_1$ was 1.20 ± 0.33 L, the mean value of $FEV_1$ (% predicted) was 51.55 ± 12.58, the mean value of $PaO_2$ was 70.57 ± 6.62 mmHg and the mean value of $PaCO_2$ was 49.33 ± 7.38 mmHg. Thirty-four of the patients (73.9%) were current or ex-smokers at the time of recruitment, and the mean smoking load was 32.78 ± 28.70 pack-years.

**Table 1.** Bacterial culture of sputum collected at stable and acute exacerbation periods.

| Microorganisms in sputum | | Patients in stable phase ($n$ = 46) | Patients at exacerbation period ($n$ = 46) |
|---|---|---|---|
| Bacterial count >$10^6$ $n$ (%) | | 15 | 21 |
| | *Haemophilus influenzae** | 7 (46.7) | 10 (47.6) |
| | *Streptococcus pneumoniae** | 3 (20.0) | 3 (14.3) |
| | *Moraxella catarrhalis** | 2 (13.3) | 1 (4.8) |
| | *Parahaemophilus influenzae** | 1 (6.7) | 2 (9.5) |
| | *Pseudomonas aeruginosa** | 2 (13.3) | 3 (14.3) |
| | Others† | 0 (0) | 2 (9.5) |

*Percentage of positive bacterial culture.

†Other microorganisms at visit 2 include one case of *Klebsiella pneumoniae* and one case of *Staphylococcus aureus*, respectively.

### 2.2. *Sputum bacterial loads and isolates*

Microbiological results on sputum at visits 1 and 2 indicated that at the stable stage 15 of the 46 sputum samples (32.6%) yielded a positive culture of potentially pathogenic microorganisms based on bacteria load >$10^6$ CFU/ml. At visit 2, the pathogens were recovered in 21 of 46 (45.65%) subjects with COPD. The dominant pathogen recovered at both visit 1 and visit 2 was *Haemophilus influenzae*; other pathogens included *Streptococcus pneumoniae*, *Branhamella catarrhalis*, *Haemophilus parainfluenzae* and *Pseudomonas aeruginosa*. The mean bacterial count at visit 2 was $10^{6.92 \pm 0.26}$ CFU/ml, which was significantly higher than for samples at visit 1 ($10^{6.70 \pm 0.33}$ CFU/ml; $P = 0.03$) (Table 1).

### 2.3. *BAL culture*

Only 9 and 11 patients accepted fiber-optic bronchoscopy at visit 1 and visit 2, respectively. Bacterial counts >$10^{3.0}$ CFU/ml in BAL specimens were 44.4% (4/9) and 54.5% (6/11), respectively. Microbiological results indicated that *Haemophilus influenzae* was also the predominant potentially pathogenic bacterium at both the stable and the exacerbation stage. In the stable stage group, no pathogen was found in BAL except for one patient, from whose sputum *Parahaemophilus influenzae* was cultured. At the exacerbation period of COPD, one patient whose sputum cultured no pathogen yielded *Streptococcus pneumoniae* in BAL (Table 2).

### 2.4. *The relationship between bacterial colonization and cytokine level at the stable stage*

Demographic characteristics, laboratory results and cytokine levels of patients with and without LABC are shown in Table 3. There were no statistical differences between two groups regarding age, BMI, pulmonary obstruction and arterial blood gas results. Sputum IL-6 and TNF-α were found to be higher in the COPD patients with LABC than in the COPD patients without LABC ($P < 0.05$). Bacterial colonization was significantly positively associated with IL-8, but was not associated with other parameters, such as the smoking habit, airway obstruction, IL-6 and TNF-α (Table 4).

**Table 2.** Bacterial culture results on BAL collected at stable and acute exacerbation periods.

| Microorganisms in BAL | Patients in stable phase ($n = 9$) | Patients at exacerbation period ($n = 11$) |
|---|---|---|
| Bacterial count $>10^3$ $n$ (%) | 4 | 6 |
| *Haemophilus influenzae** | 3 (75.0) | 3 (50.0) |
| *Streptococcus pneumoniae* | 0 (0) | 1 (16.7) |
| *Pseudomonas aeruginosa* | 1 (25.0) | 1 (16.7) |
| Others[†] | 0 (0) | 1 (16.7) |

*Percentage of positive BAL culture.

[†]Other microorganisms at visit 2 include one case of *Klebsiella pneumoniae*.

BAL — bronchoalveolar lavage fluid.

**Table 3.** Basic properties, laboratory values and level of cytokines in stable COPD patients with or without LABC.

| | Patients with LABC ($n = 15$) | Patients without LABC ($n = 31$) | $P$ value |
|---|---|---|---|
| Male: female | 14:1 | 24:7 | |
| Age (years) | 70.60 ± 6.90 | 73.03 ± 4.98 | 0.18 |
| BMI (kg/m$^2$) | 21.03 ± 2.65 | 23.56 ± 4.69 | 0.06 |
| $FEV_1$ (L) | 1.22 ± 0.33 | 1.20 ± 0.34 | 0.87 |
| $FEV_1$ (% predicted) | 48.86 ± 11.94 | 52.83 ± 12.87 | 0.32 |
| $PaO_2$ (mmHg) | 70.40 ± 5.53 | 70.65 ± 7.17 | 0.91 |
| $PaCO_2$ (mmHg) | 51.6 ± 8.00 | 48.23 ± 6.93 | 0.15 |
| Smoking load (pack-years) | 24.43 ± 18.19 | 36.66 ± 32.15 | 0.18 |
| IL-8 (pg/ml) | 792.00 ± 520.68 | 718.71 ± 377.84 | 0.29 |
| IL-6 (pg/ml) | 188.33 ± 127.19 | 104.84 ± 133.84 | 0.02* |
| TNF-α (pg/ml) | 169.33 ± 119.86 | 100.97 ± 108.85 | 0.03 |
| Interval between the first exacerbation and the baseline | 70.07 ± 36.33 | 86.48 ± 43.08 | 0.21 |

LABC — lower airway bacterial colonization; IL-6 — interleukin 6; IL-8 — interleukin 8; TNF-α — tumor necrosis factor α.

*Shows significant difference.

## 3. Discussion

Did the COPD patients have LABC at their stable stage? Did LABC contribute to the pathogenesis of AECOPD? In contrast to the sterile airways

**Table 4.** Correlations between bacterial carriage and properties, laboratory values and level of cytokines in stable COPD patients.

| | Correlation (*r* value) | *P* value |
|---|---|---|
| Age | 0.09 | 0.75 |
| BMI ($kg/m^2$) | – 0.27 | 0.34 |
| $FEV_1$ (L) | – 0.47 | 0.08 |
| $FEV_1$ (%) | 0.02 | 0.93 |
| $PaO_2$ (mmHg) | – 0.10 | 0.71 |
| $PaCO_2$ (mmHg) | 0.46 | 0.08 |
| IL-8 (pg/ml) | 0.55 | 0.03* |
| IL-6 (pg/ml) | 0.38 | 0.16 |
| TNF-α (pg/ml) | 0.01 | 0.99 |
| Symptom-free interval | – 0.36 | 0.19 |
| Smoking load (pack-years) | – 0.16 | 0.57 |

*Shows significant difference.

in normal human beings, COPD patients with active smoking and progressive airway obstruction were identified with bacteria in the lower airways, owing to the impairment of the lung defense system (including disruption of ciliary activity, mucus hypersecretion, impaired mucociliary clearance, and dysfunction of macrophage cells and granular leukocytes), as a consequence of which microbial pathogens were able to persist in the lower airways of the COPD patients.[5] Bacteria had been isolated in a relatively high proportion of patients with clinically stable COPD, indicating the presence of LABC. LABC has been shown to increase when the smoking habit and airway obstruction worsen.[6] In our study, in order to increase the incidence of bacterial colonization, only subjects who had moderate-to-severe COPD were recruited, and the majority of them were tobacco smokers. Meanwhile, in order to minimize the risk of contamination, we used quantitative cultures of sputum and BAL samples as the gold standards. During the study, we abided by the regulation of sputum sample collection and sample transport strictly; only microorganisms with counts $>10^6$ CFU/ml in sputum and $>10^3$ CFU/ml in BAL were regarded as significant. Data indicated that according to the measurement of levels of related cytokines, a threshold concentration of $10^5$–$10^6$ CFU/ml was used to define significant growth of LABC.[7] Currently, there is still no significant definition of the threshold of LABC in BAL samples, but in bacterial pneumonia, microorganisms with counts >$10^7$ CFU/ml in sputum and >$10^4$ CFU/ml in BAL were regarded as significant.[8] Therefore, a threshold concentration of $10^3$ CFU/ml was used to define significant growth of LABC in BAL samples in this study.

We found by bronchoscopy that 4 out of 9 patients with stable COPD (44.4%) were colonized with pathogens, which was a higher rate than the results obtained by sputum analysis (32.6%). This could be explained by the fact that the majority of the subjects were reluctant to accept bronchoscopy, and among the patients willing to accept the invasive procedure most were associated with purulent sputum production.[9] The potentially pathogenic bacteria in both sputum and BAL samples were predominantly *Haemophilus influenzae* (46.7% and 75%, respectively). This was similar to those found in many prior studies of sputum and lower airway bacteriology.[4]

For AECOPD, bacteria isolated from sputum and BAL were higher than those at the stable stage. The distribution of specific bacteria isolated in our study was remarkable for a large proportion of Gram-negative bacilli, and the predominant bacterium was *Haemophilus influenzae*. This indicated that bacterial infection may be of major importance in understanding the pathogenesis of AECOPD. There was a higher bacterial titer in sputum and BAL during acute exacerbation than that during stable COPD, implying that an increased bacterial count in the lower airway of patients with LABC may cause the exacerbation episode. Interestingly, the 95% confidence interval of the duration between the baseline and the first onset of exacerbation was 70.37–94.54 days, integrated with the inclusion time. We found that cold weather might also be an important factor in the pathogenesis of COPD. Further studies investigating the relationships among cold weather, airway inflammation and bacterial proliferation should refine our understanding of the pathogenesis of bacterial exacerbation and mechanisms of recurrence.

Traditional sputum microbiologic analysis may have its limitation for being contaminated by normal naso-oropharyngeal flora. BAL could give a better assessment of small airways and the alveolar situation, and was also more reliable for assessment of airway bacterial colonization than sputum by avoiding the contamination by normal naso-oropharyngeal flora. However, many patients cannot tolerate the procedure on account of invasion. In this study, we evaluated airway colonization and infection using not only sputum culture but also quantitative identification of bacteria. At the stable stage, one patient whose sputum cultured *Parahaemophilus influenzae* yielded no pathogen in BAL, indicating that *Parahaemophilus influenzae* may belong to the normal upper airway flora. Recent data showed that the presence of bacteria in the lower airway can result in a range of important pathologic effects on the lung, including activation of host defense with release of inflammatory cytokines such as IL-8, IL-6 and TNF-α, and subsequent neutrophil recruitment.[10] Chung *et al.*[11] showed that bacterial colonization in patients with stable COPD was associated with increased levels of IL-8, IL-6 and TNF-α in sputum. In the present study, in patients with LABC at the stable stage, higher sputum levels of IL-8, IL-6 and TNF-α were observed than those in patients without LABC. The differences in IL-6 and TNF-α were significant, while no significant differences were noticed in age, lung function, arterial blood gas measurements and the smoking habit between the two groups, indicating that bacterial products might affect neutrophil migration and modulate airway inflammation independently. Meanwhile, whether airway inflammation could make the individual susceptible to experiencing acute exacerbation or not needs further investigation. In multivariate regression analysis, we found that the bacterial count demonstrated a negative relationship with $FEV_1$, the interval between the first exacerbation and baseline $PaO_2$, but a positive relationship with $PaCO_2$ and the level of cytokines. In this study we have confirmed that the bacterial count is related to the sputum level of IL-8, a neutrophil chemoattractant ($P < 0.05$). All these findings suggest a possible relationship between bacterial load and airflow obstruction, decline of $PaO_2$, heightening of $PaCO_2$ or the level of cytokines, although it is uncertain whether this observation was the cause or the effect.

In conclusion, some patients with moderate-to-severe COPD have LABC at the stable stage, and LABC and bacterial infection is recognized as an important part of the pathogenesis of AECOPD. Among the bacterial pathogens colonizing at the stable stage or infecting at the exacerbation stage, *Haemophilus influenzae* was the most frequently isolated pathogen.

Both sputum culture and bacterial count results could be associated with the degree of LABC. The colonization of the tracheobronchial tree, the increase in the levels of airway IL-8, IL-6 and TNF-α at the stable stage, as well as the increase in the bacterial count, were found to be significant at the acute exacerbation period, indicating that LABC may be an important stimulus for increase in bacterial quantity and airway inflammation. All these are involved in modulating the exacerbation frequency.

## References

1. Pauwels RA *et al.*, *Am J Respir Crit Care Med*, 2001, **46**: 798–825.
2. Adams SG *et al.*, *Chest*, 2000, **117**: 1345–1352.
3. Aaron SD *et al.*, *Am J Respir Crit Care Med*, 2001, **163**: 349–355.
4. Sethi S *et al.*, *Clin Microbiol Rev*, 2001, **14**: 336–363.
5. Prieto A, *et al.*, *Am J Respir Crit Care Med*, 2001, **163**: 1578–1583.
6. Monso E *et al.*, *Eur Respir J*, 1999, **13**: 338–342.
7. Tom M, *Am J Respir Crit Care Med*, 2003, **167**: 1090–1095.
8. Zhang XY *et al.*, *Chin J Infect Chmother*, 2003, **3**: 273–276.
9. Stockley RA *et al.*, *Chest*, 2001, **117**: 1638–1645.
10. Noguera A *et al.*, *Thorax*, 2001, **56**: 432–437.
11. Chung KF, *Eur Respir J*, 2001, **18**: 50–59.

4.2

# Gene Sequence Analysis of the Norovirus Detected in Diarrhea Specimens from Children in a Surveillance Hospital in Shanghai

Zheng Teng*, Xi Zhang, Kang-Sheng Wen, Jun-Jie Shao and Bai-Hui Zhao

*Shanghai Municipal Center for Disease Control and Prevention, Shanghai 200336, China*

**Objective.** To study the genotype and genetic identity of the norovirus in the sporadic diarrhea samples collected from a surveillance hospital in Songjiang, Shanghai, in 2006.

**Methods.** The specimens were sampled from children under five years old in the surveillance hospital. The viral RNA was amplified with RT-PCR by using the human calicivirus-specific primers, and the amplified genes were sequenced and analyzed by the genetic analyzer and the DNAstar software to complete genotyping and homology analysis.

**Results.** Homology analysis revealed that the sequences of 13 specimens were homogeneous with norovirus GII. The sequences of 2 specimens were closer to the sequences from Hamburg in 2006, the sequences of 2 specimens were closer to the sequences from Utrecht in 2006, and

*Corresponding author. Email: zteng@sh163.net

the sequences of 9 specimens were closer to the sequences from Guangxi in 2006 and Tokyo between 2004 and 2005.

**Conclusion.** There were sporadic diarrhea cases in terms of norovirus genogroup GII. The majority of the sporadic strains in this study belonged to genotype GII.4 in children in the Shanghai area, and the genetic identity was varied.

*Keywords*: norovirus; genogroup; genotype; sequence analysis.

## 1. Introduction

Noroviruses (NoVs) are members of the family Caliciviridae. They are considered one of the common causes of acute viral gastroenteritis. Infection is characterized by acute onset of nausea, vomiting, abdominal pain and diarrhea. Outbreaks and sporadic cases have occurred among children and adults with a seasonality linking autumn and winter. Fecal–oral spread is an accepted major transmission mode. More than 68% of outbreaks were caused by contaminated water or food, even up to 80%.[1] Over the last five years, the outbreaks or sporadicalness of viral diarrhea infection with NoVs often occurred both in developed countries and in developing countries. And an unusual increase in the number of NoV outbreaks was reported in Europe and the United States during the winter of 2002–2003.[2]

NoV is a small, round virion 27–35 nm in diameter. It possesses a single-strand, positive-sense, polyadenylated RNA genome of 7400–7700 nucleotides. Nucleotide sequence data for NoV strains collected over 30 years have demonstrated high levels of genetic diversity. Genogroup I (GI) and genogroup II (GII) infect humans. Over the last few years, most of the emerging NoV strains have been associated with genotype GII.4 and have had a global presence.[3]

Since NoV cannot be cultured in cells, RT-PCR is currently the most sensitive diagnostic assay in methodologies. Until recently, NoV GII RT-PCR has mostly targeted ORF1.[1,3–5] In this study, RT-PCR was used to get the virally amplified products from fecal specimens. The amplified genes were sequenced and analyzed by the genetic analyzer and the DNAstar software to complete genotyping and homology analysis.

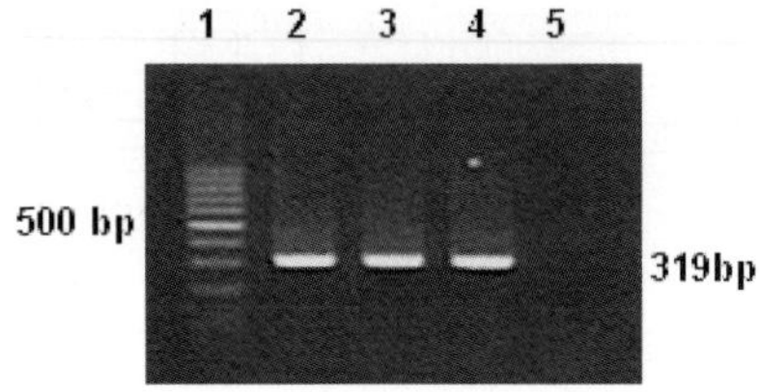


**Fig. 1.** Electrophoregram of norovirus RNA detected with RT-PCR. Lane 1 — DNA marker; lanes 2, 3 — diarrhea samples; lane 4 — positive control sample; lane 5 — negative control sample.

## 2. Results

### 2.1. *Nucleotide amplification of 15 PCR products*

319 bp was amplified from 89 specimens by one-step RT-PCR (Fig. 1).

### 2.2. *Homology search*

Fifteen NoV-positive specimens were sequenced as described above; 13 of them had nucleotide sequences. Homology searches were performed using the BLAST program. Homology analysis revealed that the sequences of 13 specimens were homologous to NoV GII.

### 2.3. *Homology analysis and phylogenetic tree construction*

NoV GII.1–GII.8 nucleotide sequences were collected according to the recommendation by China CDC. The RNA polymerase region sequences of NoV strains submitted by Japan, Germany and Holland, as well as Guangdong, Guangxi, Shandong and Hong Kong of China, between 2004 and 2006 were identified from the GenBank database. Use of the ClustalW program in the DNAstar analysis software aligned these reference sequences and 13 sequencing specimens. The homogeneity of the specimens ranged from 79.7% to 99.7% (Table 1). Phylogenetic tree construction showed that 13 sequencing specimens were constructed in 4 branches (Fig. 2).

Among all sequences of specimens aligned with the sequences deposited in the GenBank database, the sequences marked n3 and n6 were

**Table 1.** Nucleotide sequence homology among 13 norovirus strains (%).

| | N1 | N2 | N3 | N4 | N6 | N7 | N9 | N10 | N11 | N12 | N13 | N14 | N15 |
|---|---|---|---|---|---|---|---|---|---|---|---|---|---|
| N1 | — | — | — | — | — | — | — | — | — | — | — | — | — |
| N2 | 99.2 | — | — | — | — | — | — | — | — | — | — | — | — |
| N3 | 87.1 | 85.9 | — | — | — | — | — | — | — | — | — | — | — |
| N4 | 94.7 | 95.4 | 84.9 | — | — | — | — | — | — | — | — | — | — |
| N6 | 87.9 | 87.4 | 96.7 | 86.1 | — | — | — | — | — | — | — | — | — |
| N7 | 89.8 | 89.7 | 86.2 | 99.6 | 80.9 | — | — | — | — | — | — | — | — |
| N9 | 94.7 | 94.7 | 84.2 | 98.5 | 86.2 | 94.1 | — | — | — | — | — | — | — |
| N10 | 95.8 | 96.6 | 84.7 | 98.7 | 84.8 | 94.5 | 98.1 | — | — | — | — | — | — |
| N11 | 95.8 | 96.2 | 84.6 | 98.5 | 85.8 | 92.7 | 97.7 | 99.6 | — | — | — | — | — |
| N12 | 95.1 | 95.8 | 79.7 | 99.7 | 85.1 | 95.6 | 98.1 | 99.0 | 98.9 | — | — | — | — |
| N13 | 90.2 | 89.7 | 82.7 | 95.9 | 80.1 | 95.6 | 94.4 | 93.0 | 93.1 | 94.5 | — | — | — |
| N14 | 87.0 | 86.2 | 83.8 | 92.5 | 79.9 | 88.6 | 93.3 | 92.1 | 92.1 | 91.7 | 88.2 | — | — |
| N15 | 89.2 | 89.2 | 85.9 | 93.1 | 81.2 | 91.2 | 93.5 | 92.3 | 92.3 | 92.7 | 93.1 | 90.2 | — |

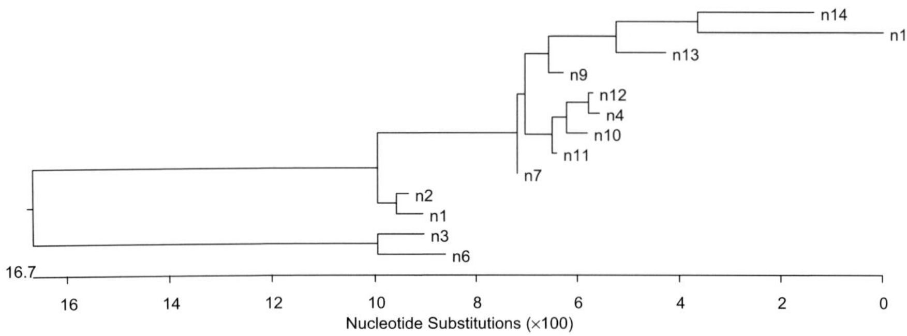


**Fig. 2.** Phylogenetic tree of the RNA polymerase region sequence of norovirus strains from children in the Shanghai area.

closer to the sequences (DQ354239 and EF200697). The nucleotide homology of these sequences was 86.2%–88.2%. The sequences marked n14 and n15 were noticeably different to the sequences from other areas. Although two sequences were near to the sequences (EF26961) on the gene tree, the nucleotide homogeneity of these sequences was about 80%. The other nine sequences aligned with the sequences (DQ026456) deposited in GenBank by Guangxi. The nucleotide identity of these nine sequences was 87.3%–97.5%, and they were closer to the sequences (DQ288303 and DQ288305) deposited by Japan between 2004 and 2005 than to those deposited by Germany (DQ1737774) and Holland (DQ440546) (Figs. 3 and 4).

## 3. Discussion

NoV is the most common etiological agent for acute gastroenteritis among children and in gastroenteritis outbreaks, and has a significant public health impact worldwide. It is difficult to cultivate NoV by cell, so establishment of a detecting method is necessary. NoV is divided into five genogroups based on the genome sequence of the RDRP and the capsid regions. NoV-GI, NoV-GII and uncertain NoV-GIV were considered to infect humans; of course, NV-GI and NV-GII were more important pathogens than NV-GIV.[3] NoV-GIII and NoV-GV were considered to infect animals. Many papers cited the primer pair 289/290 based on the

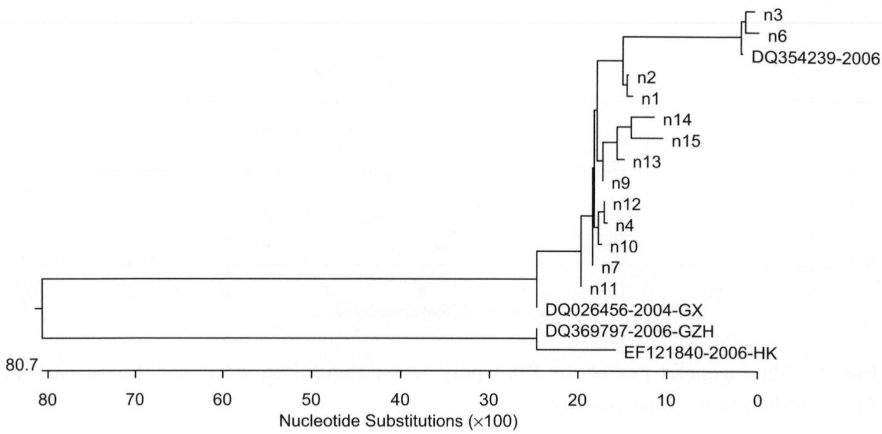


**Fig. 3.** Phylogenetic tree of the RNA polymerase region sequence of norovirus strains from China deposited in GenBank.

polymerase of 25 prototypes and currently circulating strains of HuCVs. This primer pair produces RT-PCR products of 319 bp for NoV and 331 bp for SLV.[4]

The genome of NoV is divided into three open reading frames (ORFs), polyadenylated RNA genome of 7400–7700 nucleotides. ORF1 encodes a large polyprotein. The two structural proteins — Vp1, the major capsid protein, and VP2, the minor capsid protein — are encoded by ORF2 and ORF3, respectively. Recombination of NoV is more complicated for genotyping. Twenty-three NoV recombination genotyping with an assortment of RdRp (RNA-dependent RNA polymerase) and capsid proteins and identified the ORF1-ORF2 junction as the recombination breakpoint have been identified at present.[3,5] The researchers amplified the 4865–4590 bp products, which are located at the 3′ end of RdRp.

The sequences of the sporadic diarrhea samples within the RNA polymerase region in our study showed a higher identity with the sequences in GenBank deposited by Japan in 2004 and 2005, Holland in 2006 and Germany in 2006. With alignment of the reference sequences of NoV GII.1–GII.8 within the RdRp, 11 of 13 specimens were closely related to NoV GII.4. This result is consistent with epidemiologic studies within the last three years all over the world. By homology analysis, these two

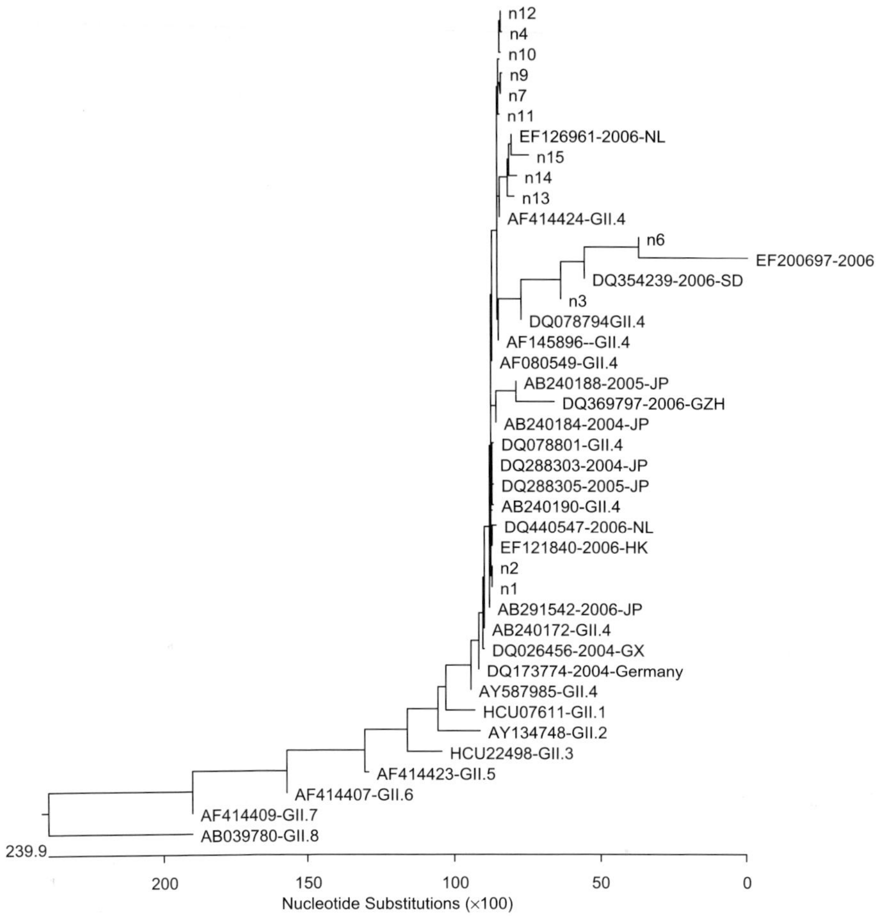


**Fig. 4.** Phylogenetic tree of the RNA polymerase region sequence of norovirus strains deposited in GenBank from 2004 to 2006.

sequences of specimens showed 87.3%, 87.0% and 84.3%, 84.2% nucleotide identity with the strain of NoV GII.4 (AB240172) and the strain of NoV GII.1 (HCU07611). As to whether sequences of two specimens are likely to vary, research is ongoing.

In publishing the alignment with sequences of Chinese strains, the sequences of specimens from Shanghai showed higher homology with those from Guangxi and Shandong provinces than from Guangdong province and Hong Kong. A phylogenetic tree was constructed based on the nucleotide sequences of the 319 bp RdRp fragment, from both the researchers' specimens and the GenBank database. The dendrogram revealed several branches, and the homology of the specimens' sequences was heterogeneous. The researchers will conduct a study to explore different ways. The researchers also discovered the sequences of 6 of 13 specimens that had over 90% homology with Guangxi strains available from shellfish living in the Bei Hai gulf. They will pay attention to the measures whether the FDA or the CDC will monitor the shell foods and the population around the infected children.

## References

1. Hedberg CW *et al.*, *Clin Microb Rev*, 1993, **6**: 199–210.
2. Dingle KE *et al.*, *J Clin Microbiol*, 2004, 3950–3957.
3. Bull RA *et al.*, *J Clin Microbiol*, 2006, **44**(2): 327–333.
4. Jiang XW *et al.*, *J Virol Meth*, 1999, **83**: 145–154.
5. Xin M *et al.*, *Chin J Virol*, 2007, **1**(23): 63–67.

4.3

# Clinical Analysis of 77 Measles Patients

Dong-Mei Shi* and Qing Xie

*Department of Infectious Diseases,*
*Ruijin Hospital, Shanghai Jiao Tong University,*
*School of Medicine, Shanghai 200025, China*

**Objective.** To examine the clinical and epidemiological features of 77 cases of measles admitted to a hospital in China in 2005.

**Methods.** A retrospective analysis was conducted on 77 cases of measles. The variables examined included age, time of year of disease onset, clinical presentation and complications of the disease, laboratory results, and prognosis.

**Results.** The peak season of measles infection was between March and June. The age of the majority of the patients ranged from 20 to 49 (94.8%). Most patients presented with a high fever, which lasted for over one week in the majority of them (63.6%). Complications included the abnormal liver function test (63.6%), pneumonia (16.9%), tonsillitis (10.9%), hematuria (8.2%), diarrhea (5.5%), stomatitis (4.1%) and palpitation (2.7%). Among them, 75 patients tested positive for measles-specific IgM antibody in their sera. The prognosis for these patients was good.

*Corresponding author: Email: sdm65@sohu.com

**Conclusions.** The peak season of measles was between March and June for primary adult and non-Shanghai residents. High fever was the major clinical feature and an abnormal liver function test was the main complication in these patients. The detection of measles-specific IgM was the key positive laboratory diagnosis test.

*Keywords*: measles; epidemiology; case study.

## 1. Introduction

Measles is an acute eruptive respiratory illness caused by the measles virus. It is commonly found in China during the spring; however, the widespread use of the measles vaccine has changed the epidemic cycle of this virus. The proportion of children who contract the disease has decreased, but adult measles has increased. In 2005, our hospital treated 77 cases of measles. In the current study, we present an analysis of the clinical features and prognosis of these 77 cases.

## 2. Clinical Data and Results

### 2.1. *General information*

During 2005, 77 cases of measles were treated by the department of infection diseases at our hospital in Shanghai, China; they were in agreement with the diagnostic criteria for measles. Out of these 77 cases, 31 were men and 46 were women; 4 ranged in age from 6 to 19 (5.2%), and 73 from 20 to 49 (94.8%). Twenty-eight cases (36.4%) occurred among the residents of the city and the remaining 49 cases were from other provinces and municipalities (63.6%). Of the 77 cases, 12 (15.6%) were infected in March, 20 (25.9%) in April, 15 (19.5%) in May and 10 (12.9%) in June.

### 2.2. *Clinical manifestations*

The following clinical symptoms were observed:

(1) Fever occurred during the eruptive stage of infection. It reached up to 40°C in some patients; there were 48 cases (62.3%) with a fever

above 39°C and 25 (32.5%) with a fever that ranged between 38°C and 39°C. Fever can last for 2–7 days.

(2) Catarrh symptoms of the upper respiratory tract appeared as conjunctival congestion, increased ocular secretions, photophobia, cough and other symptoms in 47 cases (61.0%).
(3) Koplik spots were observed in 44 cases (57.1%).
(4) One case (1.3%) of hemorrhagic rash occurred, 72 cases of congestive rash (93.5%) were observed, and 4 cases of nontypical rash (5.2%) were seen in these patients. A typical sequence of skin rash ran from the face to the trunk, then the limbs, and eventually reached the palm of the hand and the sole of the foot. The rash was described as scattered red maculopapules. It will fade when pressed. Eruption occurred after 1–2 days of fever in 16 cases (20.8%), after 3–4 days of fever in 53 cases (68.8%), and after a week of fever in 8 cases (10.4%). After the regression of the eruption, the skin began to clear, although some were observed to shed layers of skin.
(5) Other symptoms included tonsillitis (10.9%), hematuria (8.2%), diarrhea (5.5%), stomatitis (4.1%) and palpitation (2.7%).

### 2.3. *Complications*

Complications of the measles included abnormal liver function tests (63.6%) and pulmonary infection (16.9%).

### 2.4. *Prophylactic immunization*

Of the 77 patients, 63 (81.8%) reported a history of prophylactic immunization.

### 2.5. *Laboratory examination*

Laboratory examination of these patients was conducted: (1) a CBC showed that 49 cases (63.6%) displayed normal leukocytes, 24 (31.2%) displayed decreased levels of leukocytes, and 4 (5.2%) displayed increased levels of leukocytes; (2) a liver function test in 49 patients (63.6%) displayed increased ALT — 38 cases (49.4%) with ALT at 65–200 IU/L, 11 (14.3%) with ALT at 200–400 IU/L, and 3 (3.9%) with jaundice, and

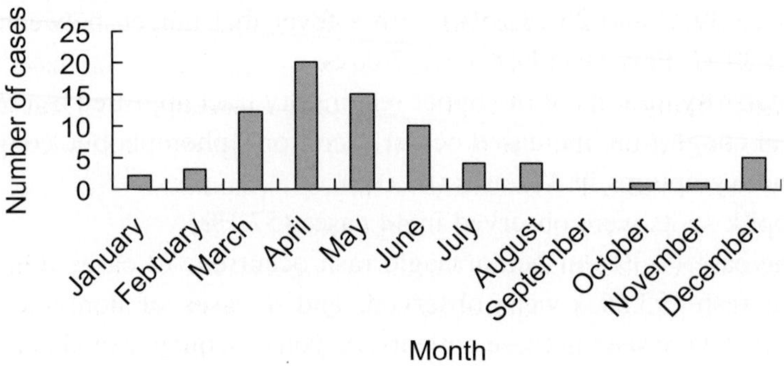


**Fig. 1.** Onset time of distribution of measles in 2005.

serum total billirubin between 25 and 34 μmol/L; and (3) the measles IgM antibody was positive in 75 of the 77 patients (97.4%) and 2 cases were negative (2.6%).

## 2.6. *Treatment and prognosis*

The patients received ribavirin, in addition to fluid and electrolytes, cooling of the skin and symptomatic treatment after hospitalization. Patients with abnormal liver function were treated with diammonium glycyrrhizinate injection of 30 ml/d, and patients with combined pulmonary infection were treated with antibiotics. Several patients had blood in the urine (+ to +++), which may have been related to the fever. The presence of blood cells in the urine disappeared after the body temperature returned to a normal level. All 77 patients recovered from the infection and were discharged. The average hospitalization without complications was 5 ± 2 days, and the average hospital day for cases with complications was 16 ± 4 days.

# 3. Discussion

Data from this analysis of patients admitted to a Shanghai hospital in 2005 showed a pattern different from the typical measles epidemic in China. This outbreak occurred from March to June, which coincided with and may have been related to the movement of the migrant population.[1] In the current population, those who had not been vaccinated against measles were more likely to be part of this migrant population; it is clear that the first step in the prevention of measles may rely on vaccination of the migrant population. Preventing the migrant population from spreading the disease has proven extremely difficult.[2]

Prior to the introduction of vaccines against measles, natural infection rates were high. During natural infection, antibodies against the virus are very effective and lifelong immunity is gained, leading to a low incidence of the disease in adults. However, immunization against measles does not lead to such a strong immune response. Therefore, in a population where many are immunized against measles, such as that of Shanghai, increased rates of infection are observed among adults. According to epidemiological investigation, the increase in adult measles is mainly due to vaccination against measles during childhood and this may be the result of decrease in measles-specific antibody levels over the years.[3] This decrease in antibody potency after inoculation may lead to infection if there is exposure to the virus.[4] During our analysis, we found that in cases where the patient had been previously immunized against measles, a more severe manifestation of the disease occurred. These symptoms included serious systemic symptoms and higher fever before rash which reached temperatures beyond 39°C and occurred in up to 62.3% of the cases. In 63.6% of the cases, accompanying alterations in liver function were observed. In general, measles patients have shown decreased appetite and deepened urine color. Elevation of liver function, as measured by ALT, is generally mild; jaundice can be observed in a small number of patients, and liver function typically returns to normal following treatment with diammonium glycyrrhizinate for 1–2 weeks. Widespread proliferation of the measles virus has been confirmed in the lymphocytes and macrophages, and is characteristic of a multinucleated giant cell with nuclear inclusions, known as the Warthin–Finkeldey giant corpuscle, which can be found mostly in the liver, spleen, tonsils, lungs, and lymph nodes.[3] We can infer that the abnormal liver function may be related to the body's immune response caused by the measles virus. Thus, it can also explain the lung inflammation, tonsillitis, and other symptoms caused by measles. In recent years, the incidence of measles pneumonia has decreased, and this may be related to the reduced incidence of children contracting measles since the main complication of childhood measles is pneumonia.

It is generally accepted that after successful vaccination with the early measles vaccine, immunity can last about 15 years.[4,5] This suggests that we should push for reimmunization with the measles vaccine to ensure that immunity against measles is maintained in order to prevent outbreaks of this disease.

## References

1. Bellini WJ *et al.*, *Lancet*, 2000, **335**: 1943–1948.
2. Yu JJ, *Immun Prog China*, 1998, **4**: 116–119.
3. Yang ZW *et al.*, *Chin J Infec Dis*, 1998, **19**(2): 84.
4. Zhang YP *et al.*, *Med Anim Contr*, 2004, **11**(20): 693–695.
5. Peng WW, *Modern Infectious Diseases and Loimology* (Science Press, 2000), pp. 733–742.

# 4.4

# Analysis of Clinical Characteristics in 77 HIV-Infected Individuals

Qi-Ming Gong*, De-Qi Qiu, Zhi-Meng Lu and Xin-Xin Zhang

*Department of Infectious Diseases, Ruijin Hospital, Shanghai Jiao Tong University, School of Medicine, Shanghai 200025, China*

**Objective.** To analyze the clinical characteristics of HIV-1 infection.

**Methods.** HIV-1-specific antibody was screened using a third generation of Abbott reagents (MEIA; S/O $\geq$ 1.00), and confirmed by the Center of Diseases Prevention and Control of Shanghai using Western blot analysis. Patient characteristics related to epidemiology, opportunistic infections and immunological statuses were analyzed.

**Results.** Seventy-seven individuals were confirmed to be HIV-positive during 1998–2006 in Rui Jin Hospital. Among them, 50.7% were Shanghai residents. The ratio of male to female was 2.85:1; this sample consisted mostly of individuals between 30 and 49 years old, and the mode of infection was sexual transmission (40.3%). The early manifestations of illness were often neglected; infection was often diagnosed as fungal infection of the digestive and pulmonary systems, *Pseudomonas aeruginosa* infection, sexual transmitted diseases (STDs), viral hepatitis, or

*Corresponding author. Email: gongqm@hotmail.com

dermatitis. Infection at multiple sites or with multiple microbes was a common characteristic of this group of patients.

**Conclusion.** Timely clinical diagnosis of HIV-1 infection is important for treatment of patients in order to prevent opportunistic infections and transmission of the virus.

*Keywords*: human immunodeficiency virus; antibody; clinical characteristics.

## 1. Introduction

The quick spread of human immunodeficiency virus type 1 (HIV-1) infection throughout the world has posed a significant medical and social challenge. As of December 31, 2007, there were 33.2 million individuals infected with HIV-1, with an increase in infected individuals more than 9000 cases per day[1]; there were a total of approximately 2.5 million new infections in 2007, Therefore, HIV-1 infection has become a very serious problem that cannot be ignored. Since the initiation of a screening test for the HIV-1 antibody in 1998, a total of 77 cases of HIV-1 infection have been detected in our hospital. Being able to identify the clinical features of HIV-1 infection may help clinicians diagnose patients with HIV-1 in early stages of the disease. In this study the clinical features of these 77 patients diagnosed with HIV-1 are summarized.

## 2. Results

### 2.1. *Epidemiological features*

#### 2.1.1. *Sample*

The sample consisted of HIV patients who were diagnosed in our hospital in the following years: 6 patients in 1998, 1 in 1999, 3 in 2000, 5 in 2001, 10 in 2002, 9 in 2003, 13 in 2004, 13 in 2005, and 17 in 2006 (Fig. 1).

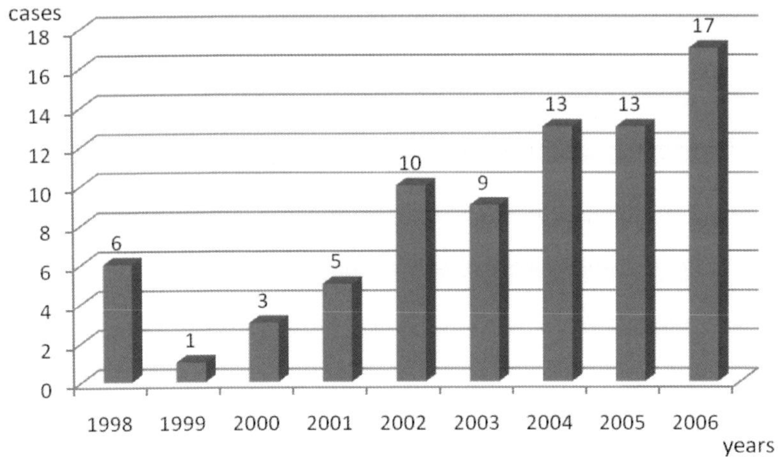


**Fig 1.** Numbers of HIV-1 infected patients from 1998 to 2006.

### 2.1.2. *Location of primary diagnosis*

Among the 77 patients, 25 were from medical wards, 5 were from surgical wards, 40 were outpatients, 5 were from emergency units, and 2 patients constituted mother-to-child transmission and were from pediatric wards.

### 2.1.3. *Age and gender distribution*

This group consisted of 57 male patients and 20 female patients; the ratio of men to women was 2.85:1. The age of the patients ranged from 4 months to 73 years, with an average of 37.5 years: 63.6% of the patients were within the 30–39 and 40–49 age groups. The age distribution of this sample is shown in Fig. 2.

### 2.1.4. *Patients' place of residence*

Three patients were foreign nationals and the remainder were from 12 provinces and cities in China, of which 39 were residents of Shanghai (50.7%) and 8 were residents of Xinjiang (10.4%).

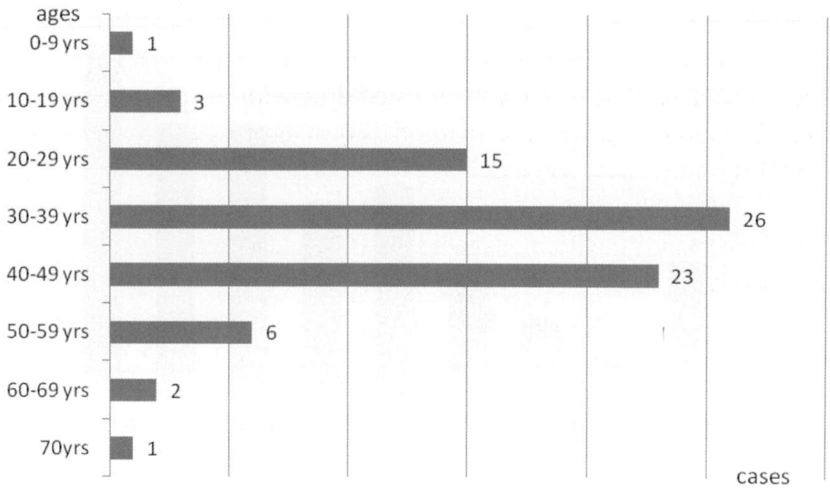


**Fig. 2.** Age distribution of patients with HIV infection.

### 2.1.5. *Mode of transmission*

Thirty-one patients (40.3%) were infected via heterosexual transmission, of which 5 had HIV-1-positive relatives. One patient may have been infected via a homosexual relationship, as there was evidence of anal infection with *Condyloma acuminata*. Blood transmission accounted for 30 patients (39.0%), of which 12 were intravenous drug abusers, 15 had a history of blood transfusion (including 7 hemophiliacs), 1 was a 13-year-old boy with a history of using unclean injection equipment, 2 were paid plasma donors and 2 were infected via mother-to-child transmission. Fourteen patients presented with an unknown mode of transmission.

## 2.2. *Opportunistic infections*

Prior to an HIV diagnosis, the first symptoms that occurred in these patients were diverse: 9 patients presented with a fever of unknown origin, 9 with a lung infection, 5 with diarrhea, 1 with nephrotic syndrome, 11 were diagnosed with a sexually transmitted disease, 7 presented with hemophilia, 5 were diagnosed with viral hepatitis, 3 with an infection of the central nervous system, 1 with a biliary tract infection, 1 with chronic cervicitis,

1 with herpes zoster, 1 with an eruption due to drug hypersensitivity, 23 had asymptomatic signs found by physical examination (of whom 5 had infected relatives), 10 were arrested drug addicts, 1 was a woman in prelabor (before delivery), 2 presented with burns, and 5 were in the hospital for a preoperation examination.

Among the 77 patients, multiple infections were present in 57 patients but not the other 20 patients (including 10 drug addicts), 5 patients with HIV-1-positive relatives, 2 patients with burns, the woman who was in the hospital for signs of prelabor, 1 patient with a drug eruption, and 1 patient with nephrotic syndrome. Among these multiple infections, there were sexually transmitted diseases (11), *Condyloma acuminate* (1), herpes zoster (1), hepatitis C (5), oral candidiasis (5), esophageal candidiasis (2), fungal diarrhea (5), pulmonary fungal infection (4), *Pseudomonas aeruginosa* pneumonia (2), *Pneumocystis pneumonia* (2), Legionnaires' infection (1), lung infection (9), *C. neoformans* infection (2), appendicitis (1), bacterial meningitis (1), biliary tract infection (1), gastric cancer (1), chronic cervicitis (1), Crohn's disease (1), pituitary tumor (1), and fever of unknown origin (9). The infections involved multiple organs, which included the lung, liver, kidney, brain, esophagus and skin.

Fifty-nine patients were infected with more than two kinds of pathogens, while 49 had more than two lesions, accounting for 83.1% of the sample; 26 patients were infected with more than three kinds of pathogens, or had more than three lesions, accounting for 44.1% of the sample; 12 patients were infected with more than four kinds of pathogens, or had more than four lesions, accounting for 20.3% of this group of patients.

## 2.3. *CD4 cell counts*

Sixteen of the 77 patients were examined for CD4 and CD8 cell counts. The number of CD4 cells ranged from 40.0 to 387.6/mm$^3$, with an average of 216.3/mm$^3$; CD8 cell counts ranged from 56.3 to 1257.1/mm$^3$, with an average of 412.4/mm$^3$; the ratio of CD4 to CD8 was 0.063 to 1.25, with an average of 0.882. Generally, bacterial infections lead to a significant increase in blood white cells as well as higher CD4 counts; however, under severe conditions, CD4 cell counts fell to a low level in patients at the end stages of the disease. In this study, statistical analysis was omitted due to the small sample size and incomplete information on some of the patients.

### 2.4. *Clinical outcomes*

Of the 77 patients, 11 died, 7 of whom had low CD4 cell counts ($< 50/mm^3$); 26 cases were lost to follow-up. Of the remaining 40 patients, 15 had highly effective antiretroviral treatment resulting in a significant increase in CD4 cell counts, and 25 are currently stable, and are continuing with follow-up care.

## 3. Discussion

Our HIV antibody testing laboratory is one of the primary HIV-1 screening laboratories authorized by the Shanghai Health Bureau and the Shanghai Municipal Center for Disease Prevention and Control. The HIV-1-positive patients from the current study were confirmed by the Shanghai Municipal Center for Disease Prevention and Control. From 1998 to 2006, the number of HIV-1-infected individuals increased on an annual basis in China. At the end of October 2006, a total of 183 733 HIV cases had been reported nationwide in China. During this period, 2313 cases were detected in Shanghai, including 718 new cases in 2006. Despite the small sample size, our results are in accordance with reports from Shanghai and also with nationwide statistics, and thus are representative of the HIV-1 epidemic in China.

There are three modes of HIV transmission: sexual transmission, mother-to-child transmission and blood transmission. It appears that the main transmission mode is sexual transmission, followed by blood transmission. Of the 77 patients in our sample, 23 (29.9%) had asymptomatic HIV infection or were detected only through physical examination prior to surgery.[2] Two cases of mother-to-child HIV transmission, one occurring in 2002 and the other in 2003, demonstrate that this route of transmission has become a real problem in this country. There have been few reports of mother-to-child HIV infections in China. The latest report issued by China's Ministry of Health and the Chinese Center for Disease Prevention and Control showed that in areas of high HIV-1 prevalence, based on the analysis of the data from over 70 000 pregnant women, the positive rate of mother-to-child transmission was 0.5%–0.7%.[3] Control of mother-to-child transmission requires active participation and cooperation among the international community, government departments, communities and families.[4]

Within this group, few HIV-1-infected patients were asymptomatic since most had one or more opportunistic infections, involving the respiratory, digestive, blood, urinary tract and nervous systems. These results suggest that the immune systems of these patients were not functioning properly. The main manifestation of HIV-1 infection is dysfunction of and a decrease in the number of CD4+ T cells. HIV-1 can infect CD4+ cells,[5] including CD4+ T lymphocytes, monocytes–macrophages, microglia cells and brain follicular dendritic cells. Replication of HIV-1 in these cells results in a decrease in the cell count and a dysfunction of these cells, leading to a deficiency in cell-mediated immunity. This deficiency in cell-mediated immunity may decrease the host's capacity of recognition and clearance of invading pathogens and immune surveillance.[6,7] Therefore, the lesions associated with HIV-1 infection are not caused by HIV-1 itself, but by the resulting immune dysfunction, which allows uncontrollable reproduction and spread of pathogens as opportunistic infections or growth of unusual tumor cells. Thus, the emergence of all kinds of opportunistic infections is an important clue to the presence of HIV-1 infection. For clinicians, analysis of the causes of various infections should be taken into consideration in combination with the patient's age, occupation, sexual activities and the history of intravenous drug use, in order to make the most appropriate judgment concerning the presence of HIV infection.

All in all, the HIV-infected patients in this study are mainly young men with low CD4+ T cell counts, often with multiple or multisite infections. In addition, there is an increasing trend of asymptomatic cases of HIV-1 infection that is only being detected through medical and laboratory examination. With the growing HIV-1 epidemic, the role of clinicians in controlling the spread of this deadly disease is critical and physicians are required to be alert and to make a timely diagnosis of HIV-1 infections.

## References

1. UNAIDS — AIDS epidemic update 2007. WHO Library Cataloguing-in-Publication Data, ISBN 9789291736218.
2. Gong QM *et al.*, *Acta Shanghai Sec Med Coll*, 2002, **22**: 432–434.
3. Wang LH *et al.*, *China Women and Children Health Care*, 2005, **20**: 350–352.
4. Li L, *China J AIDS STDs*, 2005, **11**: 316–318.
5. Cao YZ, *Diagnosis, Therapy and Care of AIDS* (People's Health Press, Beijing, 2002), pp. 18–27.
6. Brambilla AM *et al.*, *J Acquir Immune Defic Syndr*, 2001, **27**: 44–48.
7. Podlekareva D *et al.*, *J Infect Dis*, 2006, **194**: 633–641.

# 4.5

# Etiology and Clinical Characteristics of Children Hospitalized for Acute Lower Respiratory Tract Infection in Winter

Jun Shen*, Qi-Rong Zhu, Xiao-Hong Wang and Chuan-Qin Wang

*Children's Hospital of Fudan University, Shanghai 201102, China*

**Objective.** To investigate the etiology and clinical characteristics of children hospitalized for acute lower respiratory tract infection in winter.

**Methods.** A total of 381 nasopharyngeal secretion (NPS) samples from children hospitalized for acute low respiratory tract infection during the period from December 2006 to February 2007 were collected. Bacterial cultures were conducted with the NPS samples, and DFA was employed to identify seven different viruses, including respiratory syncytial virus (RSV), influenza virus (IFV) type A and type B, parainfluenza virus (PIV) type 1, 2, 3 and adenovirus (ADV) 1. The specimens were also tested for *Mycoplasma pneumoniae* and *Chlamydia pneumoniae* by polymerase chain reaction (PCR).

**Results.** Specific pathogens were identified in 381 samples, of which 81 were positive for bacterial infections, 133 for viral infections,

*Corresponding author. Email: echoshen11@yahoo.com.cn

and 12 for *Mycoplasma pneumoniae* and 6 for *Chlamydia pneumoniae*. Forty-four samples showed a mixed infection and 50.66% of the samples were positive for at least one pathogen. RSV was the predominant pathogen for three consecutive months, with 112 out of 381 specimens testing positive. Around 61.61% of the RSV cases were found in children younger than six months old and 86.61% of the cases were found in those within two years of age. Of the patients with RSV detected, 78.57% presented primarily with the wheezing symptom. Bacterial infection/colonization was demonstrated in 81 samples (21.26%). Infection with mycoplasma and chlamydia was found in 3.15% and 1.57% of the cases, respectively.

**Conclusion.** RSV remains the chief cause of acute lower respiratory tract infection among children in winter, especially infants and younger children. The etiology for more than 40% of the acute lower respiratory tract infections in children was undiagnosed in this study, indicating that a comprehensive laboratory procedure is needed to enhance etiologic diagnosis and provide effective therapies.

*Keywords*: acute lower respiratory tract infection; children; etiology; clinical characteristics.

## 1. Introduction

Acute lower respiratory tract infection (ALRTI) is the main cause of children's hospitalization. The wide application of molecular biology methods has provided tools for microbial diagnosis, but the proportion of undiagnosed respiratory infections is still as high as 20–30%. Viral infection, especially respiratory syncytial virus (RSV) infection, may be the leading cause; it is the No. 1 cause of respiratory infection in children below two years of age. We made a retrospective study on the spectrum of common causative agents for ALRTI among children in winter. The specimens collected were from the children warded in our hospital during the period from December 2006 to February 2007.

## 2. Results

### 2.1. *RSV detection*

One hundred and twelve specimens (from 112 children, respectively) were RSV-positive. Of these children, 69 (61.61%) were only 6 months old or even younger, and 97 (86.61%) were less than 2 years old. Among these 112 RSV-positive patients, 76 were boys and the boy-to-girl ratio was 2.1:1. There was no significant difference between the positive rate between boys and girls ($\chi^2 = 0.875$; $P > 0.05$), but the difference between the positive rate of six months younger (40.83%, 69/169) and older children (22.63%, 43/190) was significant ($\chi^2 = 13.80$; $P < 0.05$). The positive rate of age hierarchy is shown in Fig. 1.

The monthly positive rate from December 2006 to February 2007 was 29.73%, 30.11% and 27.38%, respectively, and the average was 29.40%. Among the RSV-positive patients, the critical illness rate was 14.29% (16 patients), while the rate was 17.00% (42 patients) among the RSV-negative patients, and statistical analysis showed no difference in the critical illness rate between the two groups. The number of RSV-positive patients who received mechanical ventilation was 3 (2.68%), versus 7 (2.83%) in the RSV-negative group. Among the 14 asthmatic patients, 4 were RSV-positive, at a ratio of 28.14%, while there were 108 patients infected with

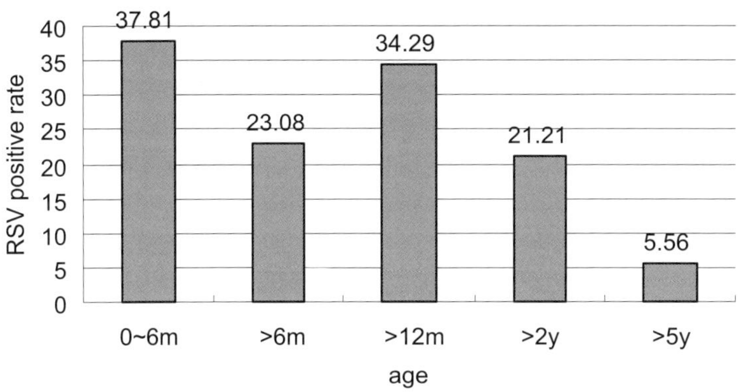


**Fig. 1.** Age distribution of 112 RSV-infected children.

RSV among the 345 nonasthmatic patients (31.30%); there was also no statistical significance ($\chi^2 = 0.047$; $P > 0.05$) between the two groups.

## 2.2. *Clinical and laboratory features of respiratory infections caused by RSV and other respiratory viruses*

See Table 1.

## 2.3. *Other respiratory pathogens detected*

Among the 359 patients, there were 4 children infected with ADV (1.11%), 7 (1.95%) with PIV type B (1 case) and type C (6 cases), and 10 (2.79%) with IFV type A (9 cases) and type B (1 case). *Escherichia coli* was found in 16 specimens, *Streptococcus pneumoniae* in 14 samples,

**Table 1.** Clinical features of respiratory infections caused by RSV and other respiratory viruses.

| Characteristics | RSV | IFV | PIV | ADV |
|---|---|---|---|---|
| Age (years) | 0.83 ± 0.84 | 2.66 ± 1.03 | 1.39 ± 1.29 | 0.98 ± 1.11 |
| Positive rate | | | | |
| Dec. 2006 | 29.73% (33/111) | 0.90% (1/111) | 1.80% (2/111) | 0.90% (1/111) |
| Jan. 2007 | 30.11% (56/186) | 3.23% (6/186) | 0.54% (1/186) | 1.08% (2/186) |
| Feb. 2007 | 27.38% (23/84) | 3.57% (3/84) | 4.76% (4/84) | 1.19% (1/84) |
| Fever | | | | |
| < 37.8 | 43.75% (49/112) | 0 | 42.86% (3/7) | 50.00% (2/4) |
| ~38.5 | 9.82% (11/112) | 0 | 14.29% (1/7) | 25.00% (1/4) |
| ~39 | 19.64% (22/112) | 30% (3/10) | 14.29% (1/7) | 0 |
| > 39 | 26.79% (30/112) | 70% (7/10) | 28.57% (2/7) | 25.00% (1/4) |
| Duration of fever (days) | 3.22 ± 3.54 | 8.9 ± 3.52 | 2.86 ± 2.98 | 2.00 ± 2.00 |
| Cough | 100% (112/112) | 100% (10/10) | 85.71% (6/7) | 100% (4/4) |
| Wheezing | 80.36% (90/112) | 40.00% (4/10) | 57.14% (4/7) | 75.00% (3/4) |
| Cyanosis | 29.46% (33/112) | 0 | 0 | 25.00% (1/4) |
| Diarrhea | 42.86% (48/112) | 20.00% (2/10) | 28.57% (2/7) | 100% (4/4) |
| WBC/$mm^3$ | 10.41 ± 2.87 | 7.15 ± 2.56 | 9.53 ± 2.92 | 11.05 ± 2.90 |
| Hospital stay (days) | 11.15 ± 5.14 | 12.20 ± 5.92 | 10.43 ± 6.61 | 8.00 ± 3.33 |
| Hospital cost (yuan) | 5975.34 ± 3585.71 | 5231.49 ± 2837.79 | 5939.06 ± 3817.73 | 3603.66 ± 1325.95 |

*Klebsiella pneumoniae* and *Staphylococcus aureus* in 10 samples, respectively, *Branhamella catarrhalis* in 8 samples and *Haemophilus influenzae* and *Haemophilus paraphrophilus* in 6 samples, respectively. Other bacterial infections included *Klebsiella pneumoniae* (4 cases), coagulase-negative staphylococcus (3 cases), *Enterobacter cloacae* (2 cases), *Acinetobacter baumannii* (2 cases), *Pseudomonas aeruginosa* (2 cases), *Klebsiella oxytoca* (1 case) and *Flavimonas oryzihabitans* (1 case). Other agents were mycoplasma (12 cases) and chlamydia (6 cases).

## 2.4. *Mixed infection*

In 44 out of a total of 381 samples, we detected at least two pathogens, and the mixed infection rate was as high as 11.55%. Among the 44 samples, 32 showed RSV coinfection with other agents. There were 5 bacteria–bacteria coinfections in all samples, including *Streptococcus pneumoniae* and *Haemophilus paraphrophilus, Streptococcus pneumoniae* and *Branhamella catarrhalis*, *Staphylococcus aureus* and *Haemophilus paraphrophilus*, *Staphylococcus aureus* and *Branhamella catarrhalis*, *Pseudomonas aeruginosa* and *Acinetobacter baumannii.* Mycoplasma was found to coexist with *Streptococcus pneumoniae*, RSV or *Staphylococcus aureus* in 3 specimens. Chlamydia was found to coexist with *Haemophilus influenzae*, type A IFV or RSV and *Enterobacter cloacae* in 3 samples.

## 2.5. *Detection rates in NPS*

Among the 381 samples, 81 (21.26%) were positive by bacterial culture, 133 (34.91%) were diagnosed as virus-positive, 12 (3.15%) mycoplasma-positive and 6 (1.57%) chlamydia-positive. Forty-four samples showed coexisting pathogens. Taken together, microbes were detected in 193 (50.66%) of 381 specimens.

## 2.6. *Patients' basic diseases*

One hundred and sixty of the 359 patients had various basic diseases. Among them, there were 45 (44.57%) children with congenital heart disease, 29 children (< 2 years old) with a history of premature delivery,

14 children with bronchial asthma, 9 children with hypersensitivity versus fungi or egg or milk, 8 with infantile hepatitis syndrome, 8 with epilepsy, 6 with anemia, 5 with gastroesophageal reflux, 4 with biliary atresia, 2 with interrespiratory infection, 2 with malnutrition, and 30 children with other basic diseases. Among the specimens from the 160 patients, 46 were diagnosed as RSV-positive, and IFV (7 cases), PIV (3 cases), ADV (1 case), mycoplasma (3 cases) and chlamydia (3 cases) were also detected. Forty-seven samples sent for bacterial culture showed positive results. The total pathogen positive rate among this population was 55.00% (88/160).

## 3. Discussion

ALRTI is the major cause of children's hospitalization. The application of molecular methods has provided useful tools for etiologic diagnosis, but still there are 20–30% respiratory infection cases in which the etiology is unidentified. Viral infection, especially RSV infection, may be the leading cause of ALRTI. RSV is also the main pathogen causing RTI in children below two years of age. It has been reported recently that human metapneumovirus (hMPV), human rhinovirus (hRV), human bocavirus (hBoV) and coronaviruses (NL263, CE43) are also important agents[1–3] in children's acute respiratory infection, which is rarely reported in China. Our study indicates that RSV is still the most important pathogen in acute asthmatic respiratory infection in children, especially among the population of children below two years of age. Besides, at least one-third of ALRTI cases in infants less than six months old may be caused by RSV. It was previously reported that RSV was also the main agent for severe respiratory infection in patients who needed intensive care or machine ventilation. Our study indicates that there was no difference between RSV-infected and noninfected patients in the rate of critical illness or the need for machine ventilation, and also that there was no difference in RSV positive rates between asthmatic and nonasthmatic children. However, the number of specimens in this study was limited. The conclusions need to be confirmed by testing more samples. RSV-infected patients are prone to manifest as asthma (80.36%) and cyanochroia (29.46%), and are often accompanied by diarrhea (42.86%). In this study the peripheral blood test, hospitalization days and the cost of hospitalization showed no difference between RSV and other virus-infected patients, but the average age of children hospitalized for RSV-infection-related LRTI was obviously lower than that of other virus-infected patients.

The positive rates of six commonly encountered respiratory viruses — ADV, type A and type B IFV, type 1, 2, 3 PIV — were only 1–3%, much less than that of RSV, which is in accordance with Wang's report.[4] The age of patients infected with these viruses was 1–3 years, indicating that infants and younger children are more susceptible to RSV infection and older children may have already been immunized through subclinical infections. Among the 381 samples, the positive rate of the 7 commonly seen viruses was 34.91%, indicating that virus infection was the major reason for children's hospitalization for ALRTI. Since we did not carry out a wider range virus spectrum screening, hMPV and hRV may have been missed in the etiologic diagnosis.

For bacterial respiratory infections, *Streptococcus pneumoniae*, *Haemophilus paraphrophilus*, *Branhamella catarrhalis* and *Klebsiella pneumoniae* were the major causes. In all the 381 samples, bacterial culture showed a positive rate of 21.26%, which was also in accordance with Wang's report.[4] The relatively high positive rate of *E. coli* and *Klebsiella pneumoniae* may correlate with the nosocomial infections; a large number of sick children had poor immunity because of their basic diseases. Owing to the extensive application of antibiotics since their hospitalization, when the children were transferred or hospitalized in our section, a much lower bacteria culture positive rate was observed. Though molecular methods are extremely sensitive in identifying the bacterial genomes in the respiratory sections, they might yield false positive results. Taking advantage of the fact that the specific and the conserved region both localized in 16S rRNA, researchers[5] detected the existence of bacteria by detecting the conserved region and differentiated pathogens through their specific region. Radstrom[6] applied the two-step method for fast detection of *Haemophilus paraphrophilus*, *Streptococcus pneumoniae* and meningococcus, and found that it had good specificity (96%) and sensitivity (94%). Specific gene probes aiming at *Haemophilus paraphrophilus*' outer membrane protein P6 gene and *Streptococcus pneumoniae*'s autolysin lyt encoding gene, together with the 16S rRNA gene as reference, could be used for fast detection of pathogens in sputum and prompt selection of antibiotics.[7] These are the directions for us to take in future studies.

The positive rate of mycoplasma and of chlamydia were 3.15% and 1.57%, respectively. The positive rate of mycoplasma was rather low compared with other studies,[4] which could be related to the lower average age of patients (1.30 ± 1.30) in our study.

Improvement in detection of causative agents during respiratory infection contributes tremendously to helping clinicians produce a timely and correct strategy for treatment. For instance, antiviral treatment (ribavirin) is effective only in the early phase of viral infections, and thus rapid and accurate diagnosis is very important, especially for intensive care unit RTI patients. In our hospital, there are still more than 40% infections undiagnosed. Therefore, long-term pathogen surveillance of pathogens, and adoption of new and more advanced technologies for this purpose, will be important for improving the quality of clinical practices.

## References

1. van den Hoogen BG *et al.*, *Nat Med*, 2001, **7**: 719–724.
2. Choi EH *et al.*, *Clin Infect Dis*, 2006, **43**: 585–592.
3. Arden KE *et al.*, *J Med Virol*, 2006, **78**: 1232–1240.
4. Wang LB *et al.*, *Chin J Infect Chemother*, 2005, **5**: 218–221.
5. Ordas J *et al.*, *J Clin Microbiol*, 2006, **44**: 2739–2742.
6. Radstrom P *et al.*, *J Clin Microbiol*, 1994, **32**: 2738–2744.
7. Fan HZ *et al.*, *J First Mil Med Univ*, 2005, **25**: 1503–1506.

4.6

# Infectious Disease Risks Associated with the Sichuan Earthquake: Experiences in Treatment of Earthquake Wound Infections

Xin Zhao*, Wen-Sheng Xu†, Jun-Xue Wang†, Xiao-Hui Miao†,‡ and Yu-Dong Gu*,‡

**Department of Hand Surgery, Huashan Hospital, Fudan University, Shanghai 200031, China*
*†Department of Infectious Diseases, Changzheng Hospital, Second Military Medical University, Shanghai 200040, China*

**Objective.** To present risks of earthquake wound-related infections and experiences in their treatment.

**Methods.** Infectious diseases reportable in China were compared in Sichuan before and after the earthquake on May 12, 2008. Wound infections were separately listed. A case report on experiences in diagnosis and treatment of earthquake wound infections is presented.

**Results.** According to the reports of the CDC from May 12, 2008 to June 30, 2008 in 18 disaster areas, 38 infectious diseases are listed as reportable infectious diseases by law. Among the wound infections related to the earthquake, 57 cases of gas gangrene have been reported.

‡Corresponding authors. Email: guyudong@fudan.edu.cn, xhmiao@infectcma.org.cn

Experiences in treatment of wound infections in four cases with multiple injuries are reported, and one special case is reported in detail.

**Conclusion.** In earthquakes, wound infections are the first priority among all infections, while respiratory and enteric infections usually occur at a later stage. Collaboration among surgeons, physicians and scientists is important for timely treatment of wound infections, and experiences gained should be summarized and disseminated.

*Keywords*: earthquake; wound infections; multiple injuries; gas gangrene.

## 1. Introduction

Earthquakes are among the most destructive natural disasters. Earthquake-related infections are variable in different stages after an earthquake.[1,2] In the immediate postdisaster stage, infections related to wounds and injuries occur at high frequencies.[3–5] During the short-term period after the earthquake, the major infectious diseases are communicable diseases, such as foodborne infections, respiratory infections and vector-borne diseases. The infectious disease spectrum and burden change, usually dramatically, after a large earthquake.

On May 12, 2008, a magnitude 8.0 earthquake struck 92 km northwest of Sichuan's provincial capital, Chengdu, in China. The area affected by this earthquake is vast, including 8 provinces and 852 counties with a total population of 348 million. An estimated 5 million buildings have collapsed, and more than 21 million buildings have been damaged in the earthquake-affected areas. The destruction and damage has resulted in massive population displacement. According to the Information Office of the State Council, as of May 26, 2008 over 65 000 people were reported dead, with the overall number of dead and missing people estimated to be in excess of 88 000. Over 360 000 were reported injured. Additionally, more than 83 000 injured people have been hospitalized after the earthquake.

According to the report from the Chinese Center for Disease Control and Prevention (CDC), infection diseases reportable by law in Sichuan province include AIDS, hepatitis B, hepatitis C, gonorrhea and syphilis, and so on. Among which hepatitis B has higher morbidity. Prior to the earthquake, the infectious diseases noticed in the area were: (1) intestinal diseases, including bacillary dysentery, infectious diarrhea, cholera, hepatitis A and hepatitis E; (2) respiratory diseases, including tuberculosis, mumps, measles, whooping cough, scarlet fever, rubella and chickenpox; (3) insectborne diseases, including malaria, dengue fever, epidemic encephalitis B, typhus, kala-azar, scrub typhus and tickborne encephalitis; (4) zoonosis, including hemorrhagic fever, rabies, brucellosis, leptospirosis, schistosomiasis, echinococcosis and clonorchiasis; and (5) bloodborne and sexually transmitted diseases, including hepatitis B, hepatitis C, gonorrhea, syphilis and AIDS. Recently, acute hemorrhagic conjunctivitis, hand, foot and mouth disease, and *Streptococcus suis*-related diseases have been reported as the main emerging diseases in the area.

After the earthquake, according to the reports of the CDC from May 12, 2008 to June 30, 2008 in 18 disaster areas, 38 infectious diseases were listed as reportable infectious diseases by law. Bloodborne infectious diseases and sexually transmitted diseases are among the top ones, reaching 38.39%. Compared to the average morbidity of the last three years, the reported number of infectious diseases and of deaths were 40.86% and 37.5% lower, respectively. Among the wound infections related to the earthquake, 57 cases of gas gangrene have been reported, of which 30 were diagnosed in the Chengdu Huaxi Hospital in Sichuan. We report here experiences in treatment of wound infections in four cases with multiple injuries suffered during the devastating Sichuan earthquake, in the Department of Hand Surgery, Huashan Hospital.

Among the four cases transferred to Huashan Hospital for unsolved problems, two presented with stable states of the illness, and obtained significant functional restoration of major injuries after a two-month treatment and returned to Sichuan to receive further rehabilitative treatment. The third patient, who suffered extensive trauma to soft tissues of the forearm and multiple fractures of the left radius and ulna, also obtained stabilization of the injuries after a series of procedures including surgical debridement, skin grafting, and topical application of traditional Chinese herbal medicine. The fourth case underwent complications of the illness and was difficult to manage during treatment, and thus is reported here in detail.

## 2. Case Report

The patient was a 40-year-old female who had been buried in a ruined building of her home on May 12, 2008, the day of the earthquake. She was evacuated 48 hours after being buried and was sent to the local hospital. On initial evaluation at the emergency department, her vital signs were detectable. She was diagnosed as having multiple crush injuries and brachial plexus injury, and received a series of treatments for two weeks in the hospital. After her vital signs became stable, she was transferred to Huashan Hospital for further treatment of the brachial plexus injury. On admission, her vital signs were stable, there were many wounds from crushing injuries all over the body surface, with scabs except for the wound on her left buttock. On the right buttock was an area of necrotic skin about 10 by 15 cm with purulent discharge, and the surrounding skin was deep red. The active movement of her left shoulder and elbow was normal, with the muscle strength grading being 3°–4° on the Medical Research Council (MRC) scale. The wrist was retained in flexion, with both active and passive extension of the wrist restricted to the same movement range. Finger flexion was normal, with the muscle strength grading being 3° on the MRC scale. Both active and passive extension of the fingers were also restricted to the same movement range, with the muscle strength grading being 3° on the MRC scale. Decreased sensation was noted in the five fingers of her left hand.

Diagnoses at the time of admission included: (1) multiple crush injuries to superficial soft tissues, with wounds in the late phase of wound healing; (2) large range necrosis of the skin and soft tissues on the right buttock, with secondary wound infection; and (3) contractures of the flexors secondary to compartment syndrome of the left forearm, with damage to the left median nerve, the ulnar nerve, and the radial nerve in the compartment. Treatment plans included, first, surgical removal of the large range necrosis of the skin and soft tissues on the right buttock to control infection, and then treatment of the flexors and the nerves in the left forearm.

A surgical operation was performed on May 28, 2008. Intraoperatively, cleaning and debridement of the necrotic tissues on the right buttock was carried out, and a local skin flap from a neighboring area was then transferred to cover the wound. The wound was packed with a dressing, and drainage equipment in negative pressure cefuroxime was given intravenously. In the following three days, the body temperature remained at 38°C, with a gradually increase in the leukocyte count and the percentage of neutrophils. Tenderness at the wound increased, and erythema and swelling around the wound at the crush injury site became evident, with production of yellow mucous secretions. *Morganella morganii* was isolated from the wound secretions. On the fourth day, the body temperature increased to 38.2°C, with chills and with a foul odor form the wound. The wound was therefore opened by removing the stitches and purulent bloody drainage from the wound was noted. The wound was irrigated with hydrogen peroxide, and this was followed by the application of a moist dressing wetted with gentamycin twice a day. Cefuroxime was discontinued, and intravenous vancomycin and fosfomycin were initiated. Between the fourth and the seventh day after the surgical operation, the body temperature fluctuated between 38°C and 39°C, and *Bacteroides fragilis* was identified in the anaerobic blood culture.

Intravenous fosfomycin and topical gentamycin were stopped based on *in vitro* antimicrobial susceptibility profiles. Intravenous vancomycin was continued, and intravenous cefoperazone and metronidazole were added. The patient showed improvement, and the body temperature and white blood cell count returned to normal on the tenth day after surgery. Intravenous vancomycin was discontinued, while intravenous cefoperazone and metronidazole were continued. On that day, surgical debridement was reperformed to remove the mass of necrotic tissue and eliminate the irregular dead space in the deep wound. The wound was left open for drainage at the end. Body temperature and white blood cell count returned to normal, and wound drainage decreased, with fresh granulation tissue forming.

## 3. Discussion

### 3.1. *Re-recognize wounds caused by crush injuries*

There has been a great deal of literature discussing the specificity in treatment of crush injuries of earthquake victims trapped by ruined buildings[6]; however, the characteristics of wounds at injury sites have been less well documented. We present the following characteristics of such wound infections from the above case:

(1) The necrotic area of skin is smaller, while the necrotic range of underlying deep tissues is much larger. In the surgical procedure for wound management, debridement of the necrotic tissues would leave the wound with a large deep space, with a relatively narrow overlying skin "mouth." Primary simple closure of this kind of wound would arrest drainage of wound secretions.
(2) There was no obvious boundary to distinguish between the necrotic and the viable tissues; the tissues in the transition area between the necrotic and the normal tissues were in a hypoxic condition and were deep red in color. Though not necrotic, these tissues had lost the characteristics of normal tissue in response to trauma, surgery, and suturing tension. Thus, the tensile strength, anti-infection ability and anticoagulation ability of these tissues were decreased, which predisposed them to developing infection and necrosis. In the first operation on this patient in our department, local flap transfer was done following the debridement, and necrosis of mass subcutaneous tissues resulted from these characteristics of such wounds.
(3) In wounds caused by crush injuries, hypoxia of tissues was in favor of proliferation of anaerobes. In the treatment of anaerobic infection, wound opening and free drainage were crucial in addition to use of antibiotics.

## 3.2. *Re-recognize treatment of wound and systemic infections*

Evaluations of this patient at admission to our department showed that her vital signs were stable. Though she had suffered multiple crush injuries, most of them were just superficial wounds, except for a large range necrosis of soft tissues on the right buttock and compartment syndrome of the forearm (it was once misdiagnosed as brachial plexus injury in the local hospital). There was no sign of systemic infection. In the first operation on this patient, local flap transfer was done following the debridement to close the wound. After the operation, chills and high fever, leukocytosis, and erythema, swelling, warming and tenderness at the wound indicated that local infection at the buttock had spread and caused bacteremic infection. With instructions from specialists on application of antibiotics in the hospital, antibiotics were timely and rationally given, bacteremic infection was controlled, and the patient showed significant progress. As surgeons we gained the following experience:

(1) Common culture of wound secretion can only identify aerobic bacteria; a negative result of common culture does not necessarily indicate "no bacteria." This was a misunderstanding in the preoperative preparation for the first operation.
(2) Therapeutic procedures should follow the basic surgical principles. Debridement and drainage constitute the most crucial step of successful treatment; closure of the wound should be undertaken in an asepsis situation. In this case, the wound was closed completely which consequently created a favorable circumstance for anaerobic bacteria proliferation and caused worsening of the illness.

Identifying the causative microorganism and using the antibiotics to which the causative microorganism was susceptible for a contaminated wound are necessary to guarantee safety of the operation. Otherwise, the patient would have the risk of spreading the infection. In this case, the striking lesson for us was that the local infection spread in the patient had became a systemic infection.

### 3.2.1. *Microorganisms evolve with time*

When microorganisms develop resistance to antimicrobial agents, new antimicrobial agents will be developed. In a reference book on clinical medicine published in 2006, 41 different types of antibiotics in the category of cephalosporin antibiotics were listed. Hence, both clinicians and scientists should constantly renew their knowledge by reading articles not limited to their fields, and close collaboration between surgeons, physicians and scientists should be encouraged.

## References


1. Toole MJ, Editorial, *Ann Emerg Med*, 1992, **21**: 418–420.
2. Ligon BL, *Semin Pediatr Infect Dis*, 2006, **17**(1): 36–45.
3. Wang ZG, *China J Trauma*, 2008, **24**(6): 401–404.
4. Tang ZH *et al.*, *China J Trauma*, 2008, **24**(6): 456–459.
5. Edicom Inc., *Clinical Drug Reference* 2006. (Sichuan Publishing Group and Sichuan Scientific and Technical Publishers, 2006), pp. 68–118.
6. Keven K *et al.*, *Scand J Infect Dis*, 2003, **35**(2): 110–113.

SECTION 5

# Immunology and Vaccinology

# Introduction and Comments: Exploring Host Immune Responses to Microbes

Shan Lu

After 60 years' success in using antibiotics against bacterial infections, society is increasingly threatened by drug-resistant bacteria such as methicillin-resistant *Staphylococcus aureus* (MRSA) and vancomycin-resistant enterococcus (VRE). Despite this emergence of antibiotic-resistant bacteria, new antibiotics generally have a narrow spectrum and have become increasingly costly to manufacture. There seems to be a fear that humans may be fighting an uphill battle against bacterial infections. Along similar lines, despite the development of new antiviral drugs, none of them have been able to clear chronic viral infections such as HIV, HBV and HCV. The best these agents can do is to suppress the infection for as long as possible to extend the lives of the infected at the cost of potentially selecting the next generation of viruses with increased drug resistance. Therefore, it is not surprising that there is an increasing amount of research on how to manipulate the body's immune system in order to control infections. In addition, during the last five years, a new wave of research programs have been developed focusing on vaccine development.

Two papers included in this book are devoted to two very different components of the body's immune system. Zheng and his colleagues (Chapter 5.1) focus on dendritic cells (DCs), the critical antigen-presenting component in an immune system. In this report, the authors study changes in subsets of DCs, an area of research with significant progress in recent years, in the context of syphilis infection. Syphilis is an ancient disease. It can be treated effectively with antibiotics, but it has re-emerged in China. Understanding the immunopathogenesis of this disease has both theoretical and practical implications. The third report (Chapter 5.3) analyzes the profiles of cytokines in HCV-infected patients. One technical advantage was the measurement of cytokine mRNAs in PBMCs instead of cytokines, which are hard to quantify when their levels are low. A number of interesting findings are reported, including the pre- and post-treatment profiles of these cytokines. The relationship of cytokine levels to treatment outcomes is analyzed in great detail. This study will be useful for setting relevant parameters to monitor the progress of HCV infection with or without active treatment.

Research on vaccines against two prominent viruses found in China is included in the current book. Zhao *et al.* (Chapter 5.2) examined the antigenicity and immunogenicity of peptide S240, which is the known receptor-binding domain from the S protein of SARS-CoV. While there has been a significant amount of work related to this epitope in the previous literature, these authors tested the reactivity of sera from two SARS patients against this peptide, implying that this epitope is immunogenic in SARS-infected patients. An important area of further study would be to determine if the antibody that is specific to this epitope is responsible for protection against this virus as determined with an *in vitro* neutralization assay. Instead, these authors conducted an immunization study and showed that this epitope is immunogenic by immunizing mice with a DNA vaccine expressing the epitope. This set of reagents can be useful for clinical monitoring of human sera with potential SARS-CoV infection.

The report by Xing *et al.* (Chapter 5.4) provides a very detailed summary of the use of DNA vaccines for the treatment of chronic HBV infections, including the immunogenicity of both surface (S) and core (C) antigens of HBV in both small animal and nonhuman primate models. Clearly, DNA vaccines expressing these HBV antigens were highly immunogenic in these studies. The next step is to determine whether such an approach will be effective in humans in controlling the infection and/or slowing down the progression of HBV infection.

Traditionally, immunology and vaccinology are two active areas in China's medical research. Even during the period of turmoil in the 1960s, China was still able to produce a broad range of vaccines for immunization of its citizens. In the last three decades, the field of immunology has experienced rapid expansion in the world and has become one of the most vibrant disciplines in biomedical science. Vaccine research, once considered unexciting and mainly confined to pharmaceutical companies, has regained fame as one of the centerpieces in global health. Therefore, it is exciting to see that China is making a significant commitment to the understanding of immunological mechanisms against a variety of microbes by addressing some very challenging questions in the areas of immunology and vaccinology, as reflected by the papers in this book.

# 5.1

# Detection of Dendritic Cell Subsets in Peripheral Blood and Levels of IL-12 and IFN-γ in Serum of Patients with Syphilis


Li-Xiong Zheng*,‡, Si-Ning Fang*, Fang-Juan Li*, Xiao-Hong Du*, Qiu-Sheng Tong*, Ya-Jie Zheng*, Hui Qi† and Xin-Gen Wang†

**Department of Dermatology*
†*Clinical Medical Research Center,*
*Shenzhen People's Hospital, Second Clinical Medicine College of Jinan University, Shenzhen 518020, China*



**Objective.** To investigate the role of dendritic cells (DCs), interleukin 12 (IL-12) and interferon γ (IFN-γ) in the pathogenesis of syphilis.

**Methods.** Flow cytometry was used to detect DC subsets in peripheral blood, and ELISA was used to detect the serum levels of IL-12 and IFN-γ in 23 patients with syphilis and 15 healthy controls.

**Results.** The expression of $CD11c^+$ DCs was (1.51 ± 2.04)% in peripheral blood of syphilis patients, which was significantly increased compared to that of healthy controls [(0.38 ± 0.18)%] ($P < 0.01$). The expression of $CD123^+$ DCs in peripheral blood of syphilis patients was



‡Corresponding author. Email: zhenglixiong@263.net

(0.37 ± 0.21)%, which was not significantly different from that of normal controls [(0.38 ± 0.22)%] ($P > 0.05$). The serum level of IL-12 was 52.70 ± 103.04 pg/ml, which was significantly increased compared to that of normal controls (0.98 ± 1.47 pg/ml) ($P < 0.05$), and the level of IFN-γ in syphilis patients was 6.12 ± 9.76 pg/ml, which was also significantly increased compared to that of normal controls (0.25 ± 0.18 pg/ml) ($P < 0.01$). There was no correlation between the expressions of DC subsets and the levels of IL-12 or IFN-γ ($P > 0.05$).

**Conclusion.** *Treponema pallidum* induced mainly $CD11c^+$ DCs and initiated cell-mediated immune responses. IL-12 and IFN-γ may participate in the immune processes after *Treponema pallidum* infection, and could play prominent roles in *Treponema pallidum* clearance.

*Keywords*: syphilis; dendritic cell; subset; interleukin 12; interferon-γ.

## 1. Introduction

Syphilis, caused by *Treponema pallidum* (TP), is a systemic, chronic sexually transmitted disease. In its early phase, the main clinical manifestation is mucosal damage to the skin. However, in its advanced phase, its primary characteristics are pathological damage to various systems and tissues, including the cardiovascular and the central nervous system, bone and the eye. It has been reported that the host immune response to TP is mainly cellular immunity. Dendritic cells (DCs) are professional antigen-presenting cells, and are highly effective in activating T cell proliferation and differentiation. There have been a series of related studies showing that DCs play an important role in antitumor and anti-infection responses, transplant rejection and autoimmune diseases, but their role in TP infection remains unclear. Interleukin 12 (IL-12) is mainly produced by activated macrophages, DCs and neutrophils, and functions in inflammation and immune regulation. DCs also function as a natural killer (NK) cell–stimulating factor. IL-12 stimulates T cells and NK cells to produce interferon γ (IFN-γ); it enhances NK cells, induces T-cell-mediated cytotoxicity and promotes Th1 cell responses.

In this study, various subsets of DCs in syphilis patients were investigated using flow cytometry. In addition, serum IL-12 and IFN-γ levels were monitored by ELISA. The correlation between DC subsets, and IL-12 and IFN-γ levels was analyzed to identify the role of DCs, IL-12 and IFN-γ in the pathogenesis of syphilis.

## 2. Results

### 2.1. *Detection of subtypes of dendritic cells and serum IL-12 and IFN-γ levels in the peripheral blood*

Myeloid DCs (MDC, $CD11c^+$) were significantly increased in syphilis patients [(1.51 ± 2.04)%] compared with those in the normal control group [(0.38 ± 0.18)%] ($P < 0.01$). However, no significant different was observed between the two groups for lymphatic DCs (LDC, CD123+). Furthermore, in syphilis patients both IL-12 and IFN-γ were significantly upregulated (52.70 ± 103.04 pg/ml, 6.12 ± 9.76 pg/ml) compared with those in the normal controls (0.98 ± 1.47 pg/ml, 0.25 ± 0.18 pg/ml) ($P < 0.05$ and $P < 0.01$, respectively), as shown in Table 1.

### 2.2. *Correlation between DC subsets and serum IL-12 or IFN-γ in the peripheral blood of syphilis patients*

There was no correlation between different subsets of DCs and serum IL-12 or IFN-γ levels in syphilis patients; nor was there a correlation between the percentages of CD11c+, CD123+ and serum IL-12, IFN-γ, or between IL-12 and IFN-γ (Table 2).

**Table 1.** DC subsets in peripheral blood and serum levels of IL-12 and IFN-γ in patients with syphilis.

| Group | *n* | CD11c+ (%) | CD123+ (%) | IL-12 (pg/ml) | IFN-γ (pg/ml) |
|---|---|---|---|---|---|
| Patients | 23 | 1.51 ± 2.04 | 0.37 ± 0.21 | 52.70 ± 103.04 | 6.12 ± 9.76 |
| Controls | 15 | 0.38 ± 0.18 | 0.38 ± 0.22 | 0.98 ± 1.47 | 0.25 ± 0.18 |
| *t* | | 2.11 | −0.18 | 1.93 | 2.31 |
| *P* | | 0.003 | 0.74 | 0.018 | 0.00 |

**Table 2.** Correlational analysis between the expressions of DC subsets and the levels of IL-12 and IFN-γ in 23 patients with syphilis.

| Cytokines | CD11c+ (%) | | CD123+ (%) | | IFN-γ (%) | |
|---|---|---|---|---|---|---|
| | *r* | *P* | *r* | *P* | *r* | *P* |
| IL-12 (pg/ml) | –0.081 | 0.714 | –0.219 | 0.315 | –0.032 | 0.885 |
| IFN-γ (pg/ml) | 0.013 | 0.953 | –0.178 | 0.416 | — | — |

## 3. Discussion

A significant increase of DCs in their immature state was observed, as these DCs were functional for antigen endocytosis and processing. During endocytosis of antigen or activation by certain stimulating factors, DCs will differentiate into a mature state characterized by the loss of phagocytic activity, secretion of chemokines and cytokines, and activation of naïve T cells by effective antigen presentation. According to the source and function, peripheral DCs can be divided into MDCs and LDCs. CD11c+ is the phenotypic characteristic of MDCs, whereas CD11c– or CD123+ is the phenotypic characteristic of LDCs. The functions of the two subsets are different. Peripheral pro-MDCs are activated during inflammation and differentiate into mature MDCs, and through synthesis and secretion of IL-12 they allow Th0 cells to differentiate into Th1 cells and trigger a Th1 response. LDCs do not secrete IL-12, and therefore Th0 cells differentiate into Th2 cells and a Th2 response is triggered. DC regulation of the balance of the immune system may occur through differentiation of different subsets of DC-induced Th0 cells. Th1 cells mediate cytotoxicity and delayed hypersensitivity, which are important in the destruction of intracellular pathogens; while the main function of Th2 cells is to stimulate B cell proliferation and produce antibodies, leading to effective humoral immunity.

In this study, results showed that peripheral CD11c+ DCs were higher in syphilis patients than in normal controls, while CD123+ DCs did not display a significant difference between the two groups. Since CD11c+ subsets prevailed over CD123+ subsets in TP infection, one may speculate that TP infection induced immature DCs to differentiate into mature MDCs, with a high percentage of CD11c+ cells. CD11c+ DCs could synthesize and secrete IL-12, thereby inducing Th0 cells to differentiate into Th1 cells and trigger a Th1 response, activating cellular immunity, and leading to destruction of TP; but LDCs were not increased, suggesting that the Th2 response is weak in TP infection.

Jong *et al.*[1] found that TP could be uptaken by DCs after being cocultured with DCs for 2h, thereby leading to DC maturation and IL-12 secretion, and to regulation of the immune response. Van Voorhis[2] also reported that IL-2, IFN-γ, IL-12 (p40) and IL-10 mRNAs could be detected in the rash of early syphilis, suggesting that the dominant Th1 local immune response activated phagocytes, and confirmed the hypothesis that IFN-γ-activated phagocytosis played a major role in the removal of TP.

Our study detected a significant increase in IL-12 (p40) and IFN-γ levels in syphilis patients compared to that in controls, suggesting that IL-12 and IFN-γ were involved in the immune responses to TP. These findings indicated that these cytokines functioned in the clearance of TP in the early phase. However, TP may evade immune clearance and escape to other organs, such as the central nervous system, where cell-mediated immune responses are not accessible.

CD11c+ DCs induced Th0 cells to develop into Th1 cells through synthesis and secretion of IL-12 and triggered a Th1 response. Theoretically, IL-12 and IFN-γ should positively correlate with CD11c+ DCs; however, our data were not in accordance with this prediction. In order to rule out the possibility of different phases or age effects in our results, correlative analysis was performed between IL-12 and IFN-γ, and CD11c+ DCs in 15 patients with phase II syphilis. After exclusion of differences in age and clinical course, there was no correlation between these factors, even with partial correlation analysis. Podwinska *et al.*[3] also observed that the expression of Th1 type cytokines such as IL-2, IFN and TNF peaked during early serum-positive syphilis, while no expression of Th2 type cytokine IL-10 was observed, suggesting that there might be antagonism between the two types of cytokines. Therefore, further investigation is needed to clarify whether the lack of a correlation between DC subsets and IL-12 and IFN-γ levels was due to complicated cross-talk among the cytokine networks.

## References

1. Jong LS *et al.*, *Yonsei Med J*, 2004, **45**(3): 515–522.
2. Van Voorhis WC *et al.*, *J Infect Dis*, 1996, **173**(2): 491–495.
3. Podwinska J *et al.*, *FEMS Immunol Med Microbiol*, 2000, **28**(1): 1–14.

# 5.2

# Analysis of Antigenicity and Immunogenicity of the Receptor-Binding Domain of SARS-CoV

Ping Zhao, Jin-Shan Ke, Min Liu, Wei Pan and Zhong-Tian Qi*

*Department of Microbiology,*
*Second Military Medical University,*
*Shanghai 200433, China*

**Objective.** To analyze the receptor-binding domain of SARS-CoV and to study its antigenicity and immunogenicity in mice.

**Methods.** The polypeptide S240 from the S protein of SARS-CoV, containing the receptor-binding domain for angiotensin-converting enzyme 2 (ACE2), was expressed in *E. coli*, and was identified with convalescent sera of two SARS patients. The S240 and EGFP fusion protein expression plasmids were constructed and used to inject BALB/c mice intramuscularly. Anti-S240 IgG antibodies of the mice were measured.

**Results.** S240 protein expressed in *E. coli* reacted with convalescent sera from SARS-CoV patients, and the fusion protein expression plasmid induced specific antibody immune responses in mice.

*Corresponding author. Email: qizt@smmu.edu.cn

**Conclusion.** These findings indicate that some important antigenic determinants might be located in the receptor-binding domain, and may play important roles in the induction of protective immune responses.

*Keywords*: severe acute respiratory syndrome coronavirus; spike protein; receptor-binding domain; antigenicity; immunogenicity.

## 1. Introduction

Recently, angiotensin-converting enzyme 2 (ACE2) has been identified as a functional receptor for the severe acute respiratory syndrome coronavirus (SARS-CoV), and the receptor-binding domain of viral spike protein has been located between amino acid residues 318 and 510.[1,2] Analysis of this receptor-binding domain of SARS-CoV and studies on its antigenicity and immunogenicity may be helpful for further understanding of the immunology of SARS-CoV and development of subunit SARS vaccines. In this study we analyzed the receptor-binding domain of SARS-CoV and studied its antigenicity and immunogenicity in mice.

## 2. Results

### 2.1. *Expression of S240 protein in* E. coli

The plasmid pET-S240 encodes a fusion protein about 29 kDa with six histidines at its C-terminal. The rare tRNA genes modified *E. coli* Rosetta (DE3) strain cells transformed with this recombinant plasmid expressed the recombinant protein induced by IPTG. In SDS-PAGE, recombinant Rosetta (DE3) expressed proteins could be visualized and the molecular weights were the same as predicted (Fig. 1).

### 2.2. *Antigenicity analysis of S240 protein*

Antigenicity of the S240 protein expressed in *E. coli* was determined by two different anti-SARS-CoV IgG positive sera of convalescent SARS

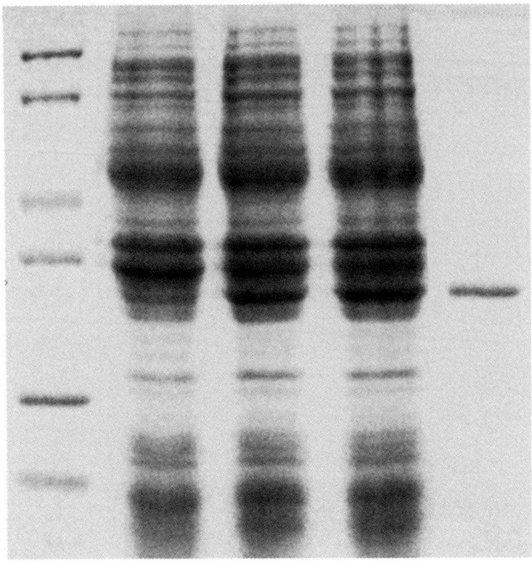


**Fig. 1.** SDS-PAGE analysis of expressed S240 protein from plasmid pET-S240–transformed *E. coli* Rosetta (DE3): (1) molecular weight marker; (2) Rosetta (DE3)/pET-S240 without IPTG induction; (3) Rosetta (DE3)/pET-S240 with 0.6 mmol/L IPTG induction; (4) Rosetta (DE3)/pET-S240 with 1 mmol/L IPTG induction; (5) purified S240 antigen using affinity resin.

patients in Western blot. As seen in Fig. 2, both of them showed specific bands as predicted, indicating that the convalescent sera of both SARS patients contained anti-S240 IgG antibodies.

## 2.3. *Antibody responses induced by the pS240-EGFP plasmid in mice*

Mouse sera were collected at 0, 2, 4, 6 and 8 weeks after the first immunization and their anti-S240 IgG antibodies were analyzed by ELISA. The anti-S240 IgG antibodies were detectable two weeks after the first immunization in the samples of the pS240-EGFP immunized group. After 1:12.5 dilution, the $A_{450}/A_{630}$ value of the mice sera in the experimental group was higher than that in the control group, and this difference was

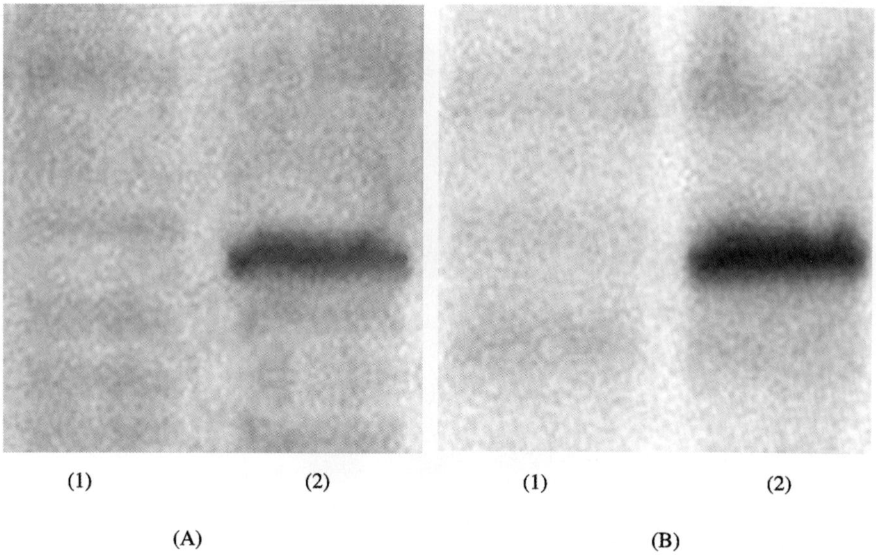


**Fig. 2.** Western blot analysis of S240 protein with two serum samples from the convalescence phase of SARS patients: (A) analysis with serum samples of SARS protein A; (B) analysis with serum samples of SARS protein B; (1) Rosetta (DE3)/pET-S240 without IPTG induction; (2) Rosetta (DE3)/pET-S240 with IPTG induction.

statistically significant ($P < 0.05$). However, none of these mice was anti-S240 IgG-positive at this time point (2 weeks). At 4, 6 and 8 weeks, the difference in the ELISA results between the experimental and control groups became more significant. And 2/4/4 of 6 mice were anti-S240 IgG–positive at 4/6/8 weeks in the experimental group (Fig. 3). The serum antibody titers of the 4 positive mice were 200, 100, 100 and 50, respectively, at 8 weeks.

## 3. Discussion

The S protein is crucial for the infectivity and pathogenicity of coronovirus. The N and C terminal parts of this protein (or the S1 and S2 subdomains of the S protein) can mediate virus–host cell binding and membrane fusion, respectively.[3] ACE2 is identified as a functional receptor for the SARS coronavirus, and the receptor-binding domain

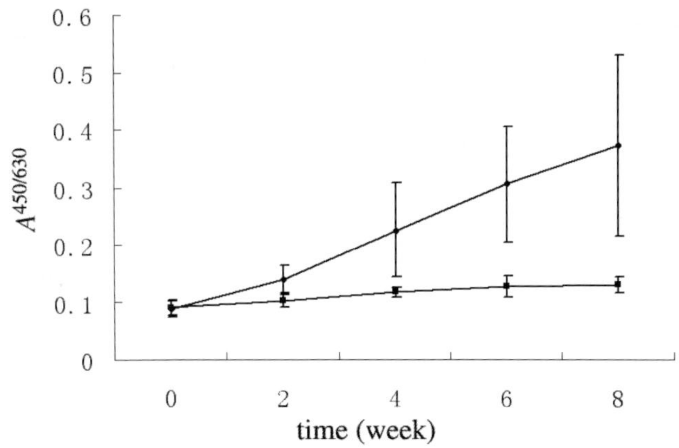


**Fig. 3.** Anti-S240 IgG in DNA immunized mice.

of viral spike protein is located between amino acid residues 318 and 510; a point mutation of Asp454 in this domain will destroy the binding ability completely.[1,2,4] Therefore, the antigenicity and immunogenicity of this receptor-binding domain should be important for the protective immune response induced by the SARS-CoV S protein.

In the present study, the polypeptide of the SARS-CoV S protein containing its receptor-binding domain was expressed in *E. coli*, and the antigenicity of this recombinant protein was determined with the convalescent sera of two SARS patients by Western blot analysis. Results indicated that both sera contained antibodies against this receptor-binding domain. It has been reported that the S protein of animal coronavirus could induce neutralizing antibody efficiently, and the protection of anti-N (or S1) antibodies was more efficient than that of anti-C (or S2) antibodies, because anti-N antibodies could block the virus–receptor interaction. In this study, the receptor-binding domain (S240) and EGFP fusion protein expression plasmid was constructed and used to inject BALB/c mice intramuscularly. By fluorescent microscopy, this fusion protein showed a long term expression in mice. Although none of the mice in the pS240-EGFP immunized group was anti-S240 IgG–positive at two weeks after the first immunization, the difference in the ELISA results between the experimental and control groups was statistically significant at this time point, indicating that the immunogenicity of this receptor-binding peptide was strong. After boosting, the difference in the antibody levels between the experimental and control groups was much more significant and four mice in the pS240-EGFP immunized group were anti-S240 IgG–positive. Though the serum antibody titers of the positive mice were not high, and the positive seroconversion rate of anti-S240 among the mice was not high, several factors may have been involved. For example, the appropriate cutoff value for antibody positivity, the detection method and the immunization schedules could all affect the results. Since the anti-S240 IgG levels in mice sera were constantly increasing during the whole observation period, by optimizing the immunization schedule antibody responses could be improved. Besides, this fusion protein was without a signal peptide and could not be secreted, which could affect its immunogenicity.

In conclusion, the receptor-binding domain of the SARS-CoV S protein showed immunogenicity in mice and reactivity to antibodies in convalescent sera of SARS patients. Further studies focusing on the neutralization effect of anti-S240 IgG antibodies on SARS-CoV and its persistence in convalescent patients with SARS may provide clues to prevention of SARS-CoV infections.

## References

1. Li W *et al.*, *Nature*, 2003, **426**: 450–454.
2. Xiao X *et al.*, *Biochem Biophys Res Commun*, 2003, **321**: 1159–1164.
3. Sanchez CM *et al.*, *J Virol*, 1999, **73**: 7607–7618.
4. Wong SK *et al.*, *J Biol Chem*, 2004, **279**: 3197–3201.

# 5.3

# Alterations in Chemokine Expression in PBMCs During and After Interferon-α Therapy in HCV Chronic Infections

Yun-Wen Hu*, Jian Wang*, Ji-Ming Zhang†, Yi Xie† and Zheng-Hong Yuan*,‡

**Key Laboratory of Medical Molecular Virology, Shanghai Medical College, Institutes of Biomedical Sciences, Fudan University, Shanghai 200032, China*
*†Huashan Hospital, Fudan University, Shanghai 200025, China*

**Objective.** To study chemokine and cytokine mRNA levels in patients with chronic hepatitis C and association with IFN therapy.

**Methods.** The mRNA levels of IL-8, T cell activation protein 3 (TCA-3 or I-309), monokine inducible by γ interferon (MIG), thymus and activation-regulated chemokine (TRAC) and macrophage-derived chemokine (MDC) in peripheral blood mononuclear cells (PBMCs) of chronic hepatitis C patients were determined by real-time PCR before and during IFN-α therapy. The ratio of chemokine/GAPDH was regarded as the relative expression level of the chemokine.

**Results.** The mRNA levels of IL-8, I-309 and MIG in peripheral blood were significantly higher in 35 patients with chronic hepatitis C than in

‡Corresponding author. Email: zhyuan@shmu.edu.cn

20 healthy donors ($P < 0.05$). During the treatment, the mRNA levels of IL-8, MIG and TARC decreased significantly. The tendency of decrease in IL-8 and MIG was similar between the pegylated interferon (PEG-IFN) plus ribavirin group ($n = 24$) and the regular interferon plus ribavirin group ($n = 11$), while TARC decreased significantly in the PEG-IFN plus ribavirin group ($P < 0.05$). The expression levels of IL-8, MIG and MDC were much higher in patients with high HCV loads (HCV RNA > 106 copies/ml; $n = 21$) than in patients with low HCV loads (HCV RNA < 106 copies/ml; $n = 14$). The mRNA levels of these five chemokines before treatment were not correlated with the serum ALT level, the HCV genotype and the outcome of the IFN therapy.

**Conclusion.** Chronic HCV infection upregulates the mRNA expression of IL-8, I-309, MIG and TARC in PBMCs, which could be improved with IFN treatment. The levels of these chemokines before treatment could not be used to predict the outcome of IFN therapy and were not correlated with the degree of inflammation.

*Keywords*: hepatitis C virus; chemokine; interferon-α.

## 1. Introduction

Hepatitis C virus (HCV) is estimated to infect more than 170 million people worldwide.[1] Of those infected, 20–50% will develop cirrhosis and hepatocellular carcinoma. Currently, interferon-α (IFN-α) is one of the most efficient drugs for the treatment of chronic hepatitis C (CHC) because of its direct antiviral effects and immune-modulating activity. While about 50% of CHC patients responded well to PEG-IFN-α plus ribavirin treatment, a proportion of the patients still exhibited no control of viral replication (nonresponders) or a relapse when therapy was stopped (relapsers).[2]

It is known that both the host and virus factors are associated with the outcome of IFN-α treatment. Viral factors, such as viral genotype and viral load, are known elements associated with a poor response to IFN-α therapy. Host factors, such as gender, race and age, play an equally important role in modulating the efficacy of IFN-α treatment. Genetic and biological factors include human leukocyte antigens (HLAs), and cytokine polymorphisms also influence therapy efficacy.[3,4] Recently it has been reported that IL-8, one member of the chemokine CXC family, could inhibit IFN-α therapy. Higher levels of IL-8 have been found in nonresponders, and the nonstructural 5A protein of HCV has been shown to induce expression of IL-8 protein.[5] However, little information is available about how the chemokine system was affected during HCV infection and associated with IFN-α therapy. In this study, we established a quantitative real-time PCR to determine mRNA levels of several cytokines and chemokines in peripheral blood mononuclear cells (PBMCs) and their association with IFN treatment in chronic hepatitis C infections.

## 2. Results

### 2.1. *Viral loads and ALT levels of CHC patients before and during IFN-α therapy*

Comparing the viral loads and ALT levels of 35 CHC patients at 0, 3 and 6 months after starting IFN-α therapy, significant decreases in the HCV RNA levels and ALT levels were found after the patients received therapy lasting 3 months (Fig. 1) (serum HCV RNA — $P_{T0\text{–}T3} < 0.001$; ALT — $P_{T0\text{–}T3} < 0.001$). But no difference was found between the 3rd and the 6th month (serum HCV RNA — $P_{T3\text{–}T6} = 0.345$; ALT — $P_{T3\text{–}T6} = 0.191$).

### 2.2. *mRNA levels of chemokines in CHC patients and the correlation with viral replication, genotypes and ALT levels*

PBMCs from 35 CHC patients before therapy and 12 healthy donors were collected and subjected to quantify the expression levels of the chemokines (Table 1). The results showed that the mRNA levels of IL-8, MIG, TARC and I-309 were significantly higher in CHC patients than in healthy donors. However, MDC expression was very low both in patients and in donors. Comparing chemokine levels between the high replication group (HCV RNA > $10^6$ copies/ml; $n = 21$) and the low replication group (HCV RNA < $10^6$ copies/ml; $n = 14$), the mRNA expressions of IL-8, MIG

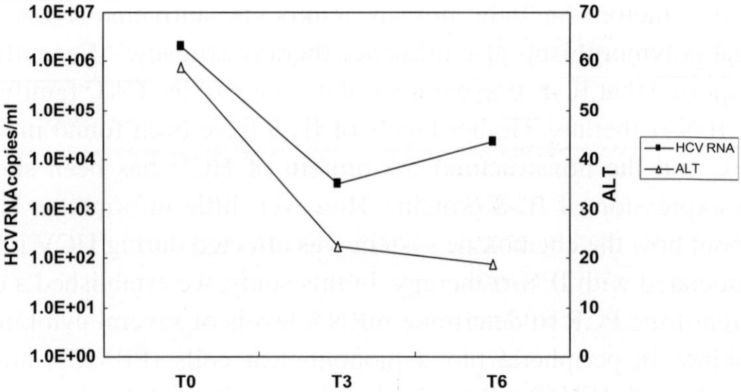


**Fig. 1.** Serum level of HCV load and ALT in chronic hepatitis C patients before and during IFN-α treatment. T0 — before IFN-α treatment; T3 — three months after starting IFN-α treatment; T6 — six months after starting IFN-α treatment.

**Table 1.** Expression levels of the chemokines in peripheral blood mononuclear cells (PBMCs) in chronic hepatitis C patients and healthy donors.

| Chemokine | Patients' mean (range) | Donors' mean (range) | *P* |
|---|---|---|---|
| IL-8 | 51.5360 (0.0416–274) | 1.2490 (0.2550–2.5100) | <0.001 |
| MIG | 0.0280 (0.0003–0.0873) | 0.0005 (0.0002–0.0014) | <0.001 |
| TARC | 0.0210 (0.0003–0.2145) | 0.0002 (0.00002–0.0005) | <0.05 |
| MDC | 0.0006 (0.00003–0.0028) | 0* | |
| I-309 | 0.0196 (0.00368–0.0979) | 0.0005 (0.0001–0.0014) | <0.001 |

*mRNA of MDC cannot be detected in healthy donors using the method in our research.

and MDC were higher in the high replication group than in the low replication group (Table 2). However, no correlation was found between the levels of these chemokines and the HCV genotype or ALT levels before treatment (data not shown).

**Table 2.** Chemokine expression level in PBMCs from chronic hepatitis C patients with high and low HCV load in serum before treatment.

| Chemokine | High HCV load group[a] | Low HCV load group[b] | *P* |
|---|---|---|---|
| | Mean (range) | Mean (range) | |
| IL-8 | 66.9271 (6.58–274) | 26.8787 (0.0416–79.7) | 0.042* |
| MIG | 0.0291 (0.0011–0.0454) | 0.0133 (0.0003–0.0873) | 0.017* |
| TARC | 0.01774 (0.0004–0.1272) | 0.02595 (0.0003–0.2145) | 0.478 |
| MDC | 0.0007 (0.0001–0.0028) | 0.0004 (0.00003–0.0015) | 0.017* |
| I-309 | 0.0223 (0.0041–0.0979) | 0.0160 (0.0037–0.0680) | 0.125 |

[a] HCV load $>10^6$ copies/ml; [b] HCV load $<10^6$ copies/ml.
$*P < 0.05$.

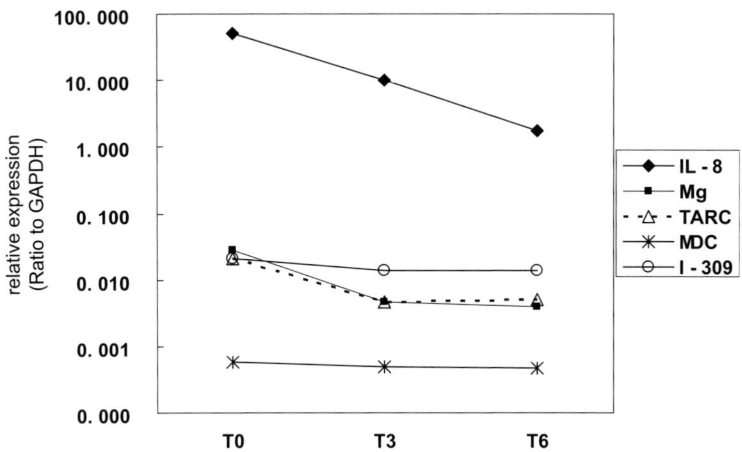


**Fig. 2.** mRNA levels of chemokines in peripheral blood mononuclear cells (PBMCs) of chronic hepatitis C patients before and during IFN-α treatment. T0 — before IFN-α treatment; T3 — three months after starting IFN-α treatment; T6 — six months after starting IFN-α treatment.

### 2.3. *Alteration of mRNA levels of chemokines during IFN-α therapy and its relation to the outcome of treatment*

Chemokine expression in PBMCs from 35 CHC patients was analyzed at 0, 3 and 6 months after starting IFN-α therapy (Fig. 2). IL-8 expression decreased continuously during 6 months of treatment. Significant decrease in the expression levels of MIG and TARC could only be found in the first 3 months of therapy. However, the levels of MDC and I-309 did not alter during treatment. Comparing the expression levels of chemokines between long-term responders ($n = 19$) and nonresponders ($n = 5$) who received PEG-IFN-α plus ribavirin therapy, no difference was found in the expression levels of IL-8, MIG, TARC, MDC and I-309 genes ($P > 0.05$).

## 3. Discussion

Recently, it has been reported that some viruses, such as human herpesvirus 6 (HHV6), cytomegalovirus (CMV) and human herpesvirus 8 (HHV8), could interact with the chemokine system to benefit persistence of viral infection.[6–10] HHV-6, CMV and HHV-8 encoded chemokine receptor-like protein, and CMV can induce several kinds of cells to secrete chemokines so as to enhance its replication. But it was not clear how HCV interacts with the chemokine system. To answer this question, we explored chemokine expression in PBMCs from CHC patients to partially discover the possible relationship between the chemokine system and HCV chronic infection.

Chemokines are a family of small cytokines, or proteins secreted by cells. The major role of chemokines is to guide the migration of cells, such as neutrophils, monocytes and lymphocytes, to sites of infection. Certain inflammatory chemokines activate cells to initiate an immune response and serve to guide cells of both the innate immune system and the adaptive immune system. Members of the chemokine family are divided into four groups (CXC, CC, C and CX3C), depending on the spacing of their first two cysteine residues. The CXC group and the CC group have more important functions. The CXC group includes IL-8, NAP2 (neutrophil-activating protein) and MIG. The CXC group mainly induces the migration of neutrophils. The CC group includes MCP-1 (monocyte chemoattractant protein 1), MIP-1α (macrophage inflammatory protein 1α), I-309, TARC and MDC. This group mainly attracts monocytes, lymphocytes and other cell types such as NK cells and dendritic cells.

IL-8 is a proinflammatory cytokine which modulates antiviral procedures through attracting neutrophils, T cells and basophils. In this study, IL-8 expression in PBMCs from CHC patients was found to be elevated before treatment and decreased following diminishment of the serum HCV load. Additionally, the IL-8 level was much higher in CHC patients with high HCV replication. However, no difference was found between the IL-8 level and the ALT level. These results suggested that IL-8 expression was directly associated with HCV replication. As Polyak *et al.*[11,12] found that HCV protein could induce secretion of IL-8 to antagonist the antiviral effect of IFN-α, we predicted that IL-8 might be utilized by HCV to establish chronic infection.

MIG/CXCL-9 is a member of the CXC chemokine family, which attracts T cells and NK cells through the chemokine receptor CXCR3. Studies have reported the antiviral and antitumor effect of MIG.[13,14] Our study found that, like IL-8, MIG expression was upregulated during HCV chronic infection and downregulated after IFN-α therapy. A significant difference existed between the high and low HCV load groups, which suggested its relationship to HCV replication. Since MIG is induced by IFN-γ, high expression of MIG in the patient's PBMCs reflected a high level of endogenous IFN-γ in hosts. It has been reported that high levels of endogenous IFN-γ could negatively modulate the antiviral efficiency of IFN-α.[15] Whether higher MIG expression was correlated with lower reaction to IFN-α treatment needs to be studied further.

I-309, which is secreted by activated T cells, belongs to the CC chemokine family. It is the ligand of CCR8 and specially attracts the Th2 cell, which expresses CCR8.[16,17] Our study showed higher expression of I-309 in CHC patients than in donors. This indicated that T cells were activated during HCV infection. Meanwhile, the I-309 expression level did not alter obviously when HCV replication was inhibited by IFN-α therapy. This suggested that T cells activated by HCV infection could persist for a long time.

Both TARC and MDC are members of the CC chemokine family. It was considered that these two proteins have similar functions, such as attracting Th2 cells to sites of infection and leading to tissue damage.[18,19] Elevated secretion of TARC and MDC was found in serum from patients with atopic dermatitis or systemic lupus erythematosus.[20,21] However, TARC and MDC were also associated with immune escape of virus. Latent membrane protein 1 (LMP1) encoded by Epstein–Barr virus could induce infected B cells to produce TARC and MDC, which could enhance Th2 response and protect Epstein–Barr virus against escaping from elimination by Th1 type immune response.[22] Our study found that the TARC level was elevated by HCV infection and decreased following decline of the serum HCV load during IFN-α therapy. This suggested positive correlativity between Th2 response and HCV replication. Enhancing Th2 response by HCV through inducing chemokines to antagonist Th1 response may be one of the mechanisms which benefit persistence of infection. Since the expression level of MDC was very low in both CHC patients and healthy donors, more sensitive methods should be used to study the exact role of MDC in HCV infection.

Above all, this study showed that the expression levels of IL-8, MIG, TARC and I-309 were changed in HCV chronic infection. The relationship between the levels of these chemokines and HCV replication could partially explain the mechanisms of persistence of infection and resistance to IFN-α therapy. But to clarify the exact role of the chemokine system in HCV chronic infection, intensive studies should be undertaken in the future, which may help to find a new drug target for antiviral treatment and improve the rate of response to IFN-α therapy.

## References

1. Alter MJ *et al.*, *N Engl J Med*, 1992, **327**: 1899–1905.
2. Fried MW *et al.*, *N Engl J Med*, 2002, **347**: 975–982.
3. Barrett S *et al.*, *J Med Virol*, 2003, **71**: 212–218.
4. Konishi I *et al.*, *Intervirology*, 2004, **47**: 114–120.
5. Khabar KS *et al.*, *J Interferon Cytokine Res*, 2002, **22**: 1005–1012.

6. Isegawa Y *et al.*, *J Virol*, 1998, **72**: 6104–6112.
7. Gao JL *et al.*, *J Biol Chem*, 1994, **269**: 28539–28542.
8. Guo HG *et al.*, *B Virology*, 1997, **228**: 371–378.
9. Khabar KSA *et al.* (Science Publishers, Enfield, N.H., 2000).
10. Osawa Y *et al.*, *Infect Immun*, 2002, **70**: 6294–6301.
11. Polyak SJ *et al.*, *J Virol,* 2001, **75**(13): 6209–6211.
12. Polyak SJ *et al.*, *J Virol*, 2001, **75**(13): 6095–6106.
13. Liu MT *et al.*, *J Immunol*, 2001, **166**: 1790–1795.
14. Kanegane C *et al.*, *J Leukoc Biol*, 1998, **64**: 384–392.
15. Thimme R *et al.*, *Proc Natl Acad Sci USA*, 2002, **99**: 15661–15668.
16. Haque NS *et al.*, *Circulation*, 2000, **102**: 786–792.
17. Iellem A *et al.*, *Eur J Immunol*, 2000, **30**: 1030–1039.
18. Mantovani A *et al.*, *J Leukoc Biol*, 2000, **68**: 400–404.
19. Shimada Y *et al.*, *J Dermatol Sci*, 2004, **34**: 201–208.
20. Morita E *et al.*, *J Dermatol Sci*, 2004, **34**: 237–240.
21. Okamoto H *et al.*, *J Rheumatol*, 2003, **30**: 2369–2373.
22. Nakayama T *et al.*, *J Virol*, 2004, **78**: 1665–1674.

# 5.4

# A Summary of Immunogenicity Studies on DNA Vaccines Expressing the Hepatitis B Virus Core and Surface Antigens as Tested in Both Murine and Nonhuman Primate Models

Yi-Ping Xing[*,†,‡], Zu-Hu Huang[*,†,‡], Shi-Xia Wang[‡,§], Jun Li[*,†,‡] and Shan Lu[*,‡,§,¶]

*Jiangsu Province Infectious Diseases Key Laboratory
†Department of Infectious Diseases
‡China–US Vaccine Research Center,
The First Affiliated Hospital, Nanjing Medical University,
Nanjing 210029, China
§Laboratory of Nucleic Acid Vaccines, Department of Medicine,
University of Massachusetts Medical School,
Worcester, MA 01605, USA

## 1. Introduction

It is estimated that there are more than 350 million chronic hepatitis B virus (HBV) carriers in the world today, and there is a significant association among persistent infection, liver cirrhosis and hepatocellular carcinoma.[3,9] Although a few antiviral drugs can effectively inhibit HBV

¶Corresponding author. Email: shan.lu@umassmed.edu

replication, they rarely eliminate the intranuclear viral covalently closed circular DNA, which is responsible for the persistence of infection. Recent data indicate that immunotherapeutic strategies, stimulating both cellular and humoral immune responses to HBV antigens, are essential for ultimately controlling chronic HBV infection. The control of HBV infection is thought to be mediated by both humoral and cellular immune responses involving protective antibodies, as well as class I and class II major histocompatibility complex (MHC)–restricted T cells.[5,6,8,20] HBcAg and HBeAg are expressed by the HBV core gene as two partially colinear antigens. During HBV infection, HBcAg and HBeAg are important targets of antiviral immunity. HBcAg and HBeAg are serologically distinct but share all characterized T cell epitopes due to the extensive amino acid homology.[7,11,18,19] During acute HBV infection, cytotoxic T lymphocytes (CTLs) specific for HBcAg and HBeAg can be detected in the circulation of the infected hosts. In contrast, in chronic HBV infection, HbcAg- and HBeAg-specific CTL and T-helper cell activity are not readily detected. The accumulated data suggest that CTL activity may play an important role in resolving HBV infection.[15,18,19]

The full length HBV surface protein (HBsAg) includes three domains: pre-S1, pre-S2 and S. Different combinations of these domains produce what are known as the small or major (S alone), middle (M, pre-S2 + S) and large (L, pre-S1 + pre-S2 + S) surface proteins, all of which are encoded in the S viral gene and exist as either unglycosylated or *N*-glycosylated forms of viral DNA in the host cell.

The nature of HBV infection calls for two different types of immune interventions to control this epidemic. The first is a prophylactic vaccination program to prevent any new infections from occurring and the second is immunotherapy for those already infected with the virus. DNA vaccination has emerged as an attractive approach to inducing balanced immune responses. DNA immunization has been shown to elicit cell-mediated immunity, as evidenced by stronger CTL responses to HBV envelope proteins when compared to a protein immunization.[1] Furthermore, activation of T cells has been shown in mice that had received a DNA plasmid encoding for the S gene of the HBV envelope in both normal and transgenic mice which constitutively express HBsAg in the liver and therefore are an indicator of the utility of this vaccine formulation as a form of therapy.[4,12] In an evaluation of the DNA prime-protein boost vaccination regimen, twice priming with a pDNA encoding HBsAg and a one-time boost with the recombinant HBsAg protein induced greater IL-12 and interferon $\gamma$ in splenocytes of DNA-vaccinated mice, plus stronger CD8 + CTL responses when compared to either vaccine modality alone.[22]

In this report, we summarize the key results of HBV DNA vaccine studies already conducted by our group to describe the overall ability of HBcAg- and HBsAg-expressing DNA vaccines to elicit humoral and cellular immune responses in both small animal and nonhuman primate models. In addition, pilot studies were conducted to analyze the differences in immune response elicited by different DNA vaccination approaches. Finally, we examined the roles of glycosylation in the immunogenicity of DNA vaccines since intracellular expression of antigens is critical for gene-based vaccines.

## 2. Results

### 2.1. *Immunogenicity of HBc DNA vaccines in small animals*

#### 2.1.1. *Expression of HBc by DNA vaccines*

An HBc gene insert was PCR-amplified from HBV Adr strain genome DNA and built into DNA vaccine vector pJW4303. Expression of HBc protein was confirmed by detection of the protein in transiently transfected 293T cells and verified by Western blot. The HBc protein had the expected molecular weight of 21 kDa and was efficiently expressed in the lysate of 293T cells transiently transfected with the HBc DNA vaccine.

#### 2.1.2. *Induction of HBc-specific antibody responses in C57BL/6 and BALB/C mice by the HBc DNA vaccine*

Both C57BL/6 and BALB/C mice (5 mice/group) were immunized with the HBc DNA vaccine via either gene gun or IM DNA immunization, using respectively a dose of 6 μg or 100 μg of DNA. HBc-antigen-specific antibody responses were detected by ELISA in immunized animals and the levels of antibody responses rose following each of the three immunizations, until they reached peak levels after the third

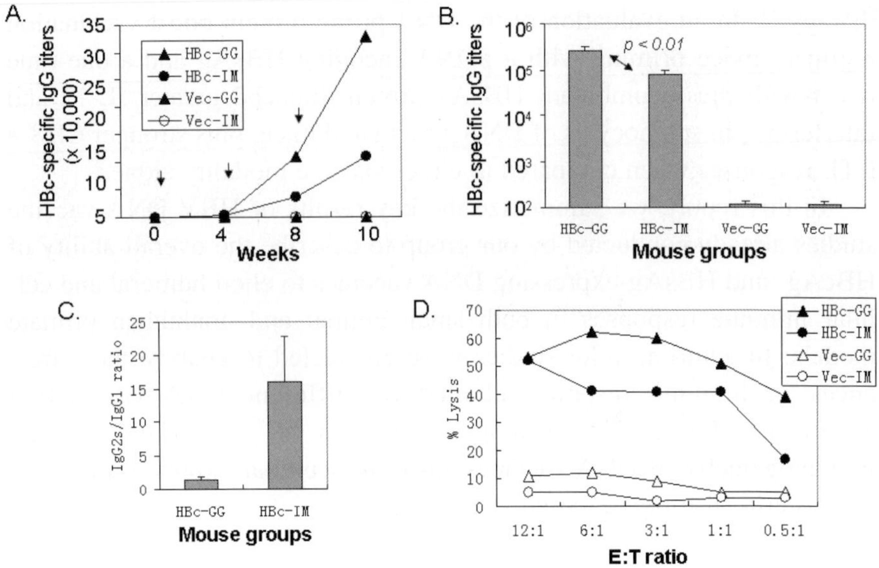


**Fig. 1.** HBc-specific antibody and CTL responses in C57/BL6 mice induced by either the HBc DNA vaccine (HBc) or the empty vector (Vec). The mice were immunized through either gene gun inoculation (GG) or intramuscular injection (IM). The antigen-specific IgG responses were measured by ELISA against HBc antigen in panels A–C. A: HBc-specific IgG titers over time in pooled mouse sera of each group; the arrows indicate the time points of DNA immunizations. B: The peak level of HBc-specific IgG titers two weeks after the third DNA immunization. The values represent mean IgG titers of 5 mice/group ± standard deviations. The *P* value indicates the statistical significance between different immunization groups. C: The analysis of HBc-specific IgG isotypes (IgG1/IgG2a ratio). The values represent average IgG1/IgG2a ratios of 5 mice/group ± standard deviations using sera collected two weeks after the third DNA immunization. D: HBc-peptide-specific CTL responses four weeks after the third DNA immunization. The ratios of effector:target cells (E:T) are shown. The % lysis is expressed as the average of 5 mice/group.

DNA immunization (Fig. 1A). High levels of anti-HBc antibodies (>1:100,000) were elicited as measured by end titration titers of peak level animal sera in both C57BL/6 and BALB/C mice (Figs. 1B and 2A). Overall, gene gun immunization in mice of both strains elicited significantly higher HBc-specific antibody titers than IM immunization (Figs. 1B and 2A), while the mouse sera immunized with the empty DNA vector did not display any HBc-specific antibody responses.

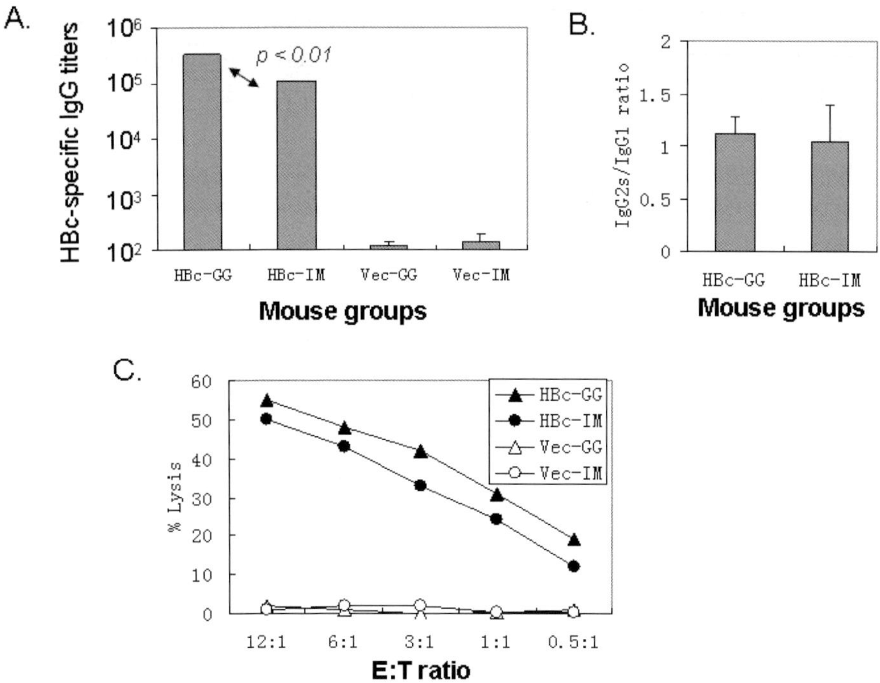


**Fig. 2.** HBc-specific antibody and CTL responses in BALB/C mice induced by the HBc DNA vaccine (HBc) or the empty vector (Vec). The mice were immunized through either gene gun inoculation (GG) or intramuscular injection (IM). The antigen-specific IgG responses were measured by ELISA against HBc antigen in graphs A and B. A: The peak level of HBc-specific IgG titers two weeks after the third DNA immunization. The values represent mean IgG titers of 5 mice/group ± standard deviations. The *P* value indicates the statistical significance between different immunization groups. B: The analysis of HBc-specific IgG isotypes (IgG1/IgG2a ratio). The values represent average IgG1/IgG2a ratios of 5 mice/group with standard deviations. The sera were collected two weeks after the third DNA immunization. C: HBc-peptide-specific CTL responses four weeks after the third DNA immunization. Levels of lysis at different ratios of effector:target cells (E:T) were analyzed. The % lysis is expressed as the average of 5 mice/group.

Further analysis of the subtypes of IgG responses showed a predominantly Th1 type antibody response, as measured by the IgG2a/IgG1 ratio (>1) in immunized C57BL/6 mouse sera by either gene gun or IM DNA immunization (Fig. 1C). However, IM DNA immunization elicited more IgG2a and more toward Th1 antibody responses. This is consistent with a previous finding that IM DNA immunization elicits more of a Th1 type immune response when compared to gene gun immunization.[11,30,38,39] In BALB/C mice, more balanced Th1/Th2 antibody responses were produced by either gene gun or IM DNA immunization with IgG2a/IgG1 ratios (~1) in BALB/C mice (Fig. 2B). These results suggest that immune responses with the same HBc DNA vaccine in different mouse strains may differ.

### 2.1.3. *Induction of HBc-specific CTL responses in C57BL/6 and BALB/C mice by the HBc DNA vaccine*

Mouse splenocytes were collected after the final HBc DNA immunization to analyze HBc-peptide-specific CTL responses. Overall, the CTL responses were very similar between two mouse strains. The HBc DNA vaccine elicited strong specific CTL responses regardless of whether the DNA was delivered via gene gun or IM immunization (Figs. 1D and 2C). The CTL background level induced by the empty DNA vector was very low (negative) in both animal strains. Compared to the control group, the specific CTL killing with the highest specific lysis reached 50%–60% at an effector:target ratio of 12:1 in the experimental group which received the HBc DNA vaccine. These results also demonstrated that mice which received the gene gun immunization were able to produce better CTL responses (Figs. 1D and 2C), further suggesting that the gene gun approach was more effective as a DNA vaccine method when compared to IM immunization.[21]

### 2.1.4. *Improved HBc-specific immune responses by the adjuvant QS-21*

Based on the results above that gene gun immunization was more effective in inducing HBc-specific responses, we tested whether the adjuvant QS-21 could further improve the HBc-specific immune responses in C57BL/6 and BALB/C mice. After 24 h of receiving HBs DNA immunization, mice were intradermally injected with 20 μg of QS-21 in 200 μg of PBS at six gene gun DNA delivery sites. Mice that received QS-21 following gene gun immunization produced significantly higher

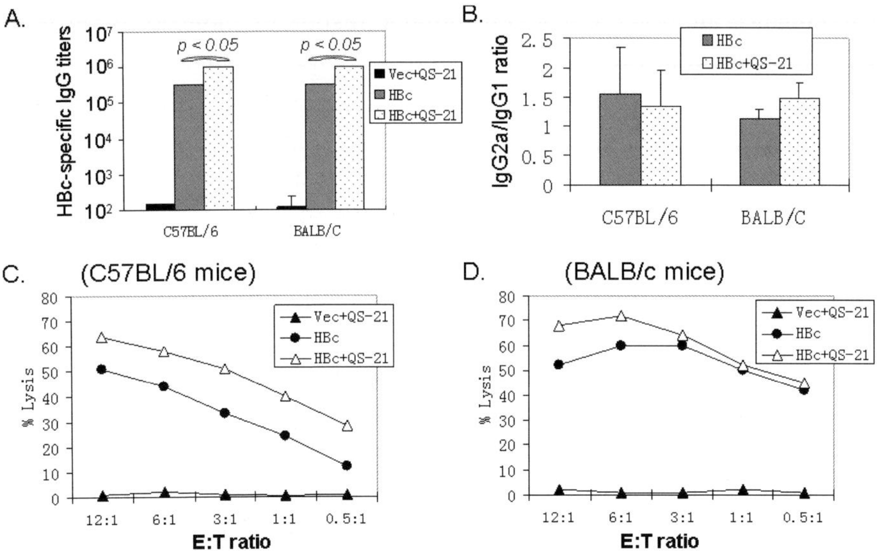


**Fig. 3.** HBc-specific antibody and CTL responses in C57/BL6 and BALB/C mice induced by the HBc DNA vaccine with or without the adjuvant QS-21. The antigen-specific IgG responses were measured by ELISA against HBc antigen in graphs A and B. A: The peak level of HBc-specific IgG titers two weeks after the third DNA immunization. The values represent mean IgG titers of 5 mice/group ± standard deviations. The *P* value indicates the statistical significance between different immunization groups. B: The analysis of HBc-specific IgG isotypes (IgG1/IgG2a ratio). The values represent average IgG1/IgG2a ratios of 5 mice/group ± standard deviations. The sera were collected two weeks after the third DNA immunization. C, D: HBc-peptide-specific CTL responses four weeks after the third DNA immunization, in C57/BL6 and BALB/C mice, respectively. Data are shown at different effector:target cell ratios. The % lysis is expressed as the average of 5 mice/group.

HBc-specific antibody titers ($P < 0.05$) after three DNA immunizations, while the HBc-specific antibody response was negative in mice receiving QS-21 with the vector alone (Fig. 3A). However, QS-21 did not change the profile of IgG subtypes, and the IgG2a/IgG1 ratios were very similar in the groups that did not receive QS-21 in both C57BL/6 and BALB/C mice. Further analysis of CTL responses showed that improved HBc-specific CTL responses were elicited by QS-21 in both C57BL/6 and BALB/C mice (Figs. 3C and 3D). This suggests that the QS-21 adjuvant can enhance antibody and T cell immune responses induced by an HBc DNA vaccine in mice.

## 2.2. Immunogenicity of HBs DNA vaccines in small animals

### 2.2.1. *Expression of HBs by DNA vaccines*

The gene insert of the HBs middle antigen (including the pre-S2 domain and small S antigen) was also PCR-amplified from HBV Adr strain genome DNA and cloned into DNA vaccine vector pSW3891. Expression of HBs protein was confirmed by detection of transient expression in 293T cells and verified by Western blot. The HBs proteins with expected molecular weights were efficiently expressed in both the supernatant and the lysate of 293T cells transiently transfected with the HBs DNA vaccine, as verified by Western blot using HBs-specific mouse antibody. There are three potential Asn (N)–linked glycosylation sites (N4, N114 and N201) in the HBs middle antigen. The N-linked glycosylation on the HBs protein expressed by the HBs DNA vaccine was further analyzed by using the specific enzyme PNGase, which served to remove the N-linked glycosylations. After the treatment with PNGase, the HBs protein expressed from the transfected 293T cell supernatant had the expected reduced molecular weight, suggesting that the HBs protein expressed by the HBs DNA vaccine is glycosylated with *N*-linked glycosylations.

### 2.2.2. *Induction of HBs-specific antibody and CTL responses by the HBs DNA vaccine in mice*

Both C57BL/6 and BALB/C mice (6 mice/group) were immunized with the HBs DNA vaccine via gene gun immunization. HBs antigen-specific antibody responses were detected by ELISA in immunized animals, and the levels of antibody responses were generated after three immunizations. High levels of anti-HBs antibodies (> 1:1000) were elicited in both C57BL/6 and BALB/C mice (Figs. 4A and 4B). Further analysis of HBs peptide-specific CTL responses demonstrated that the HBs DNA vaccine also produced high levels of specific CTL responses in mice, with

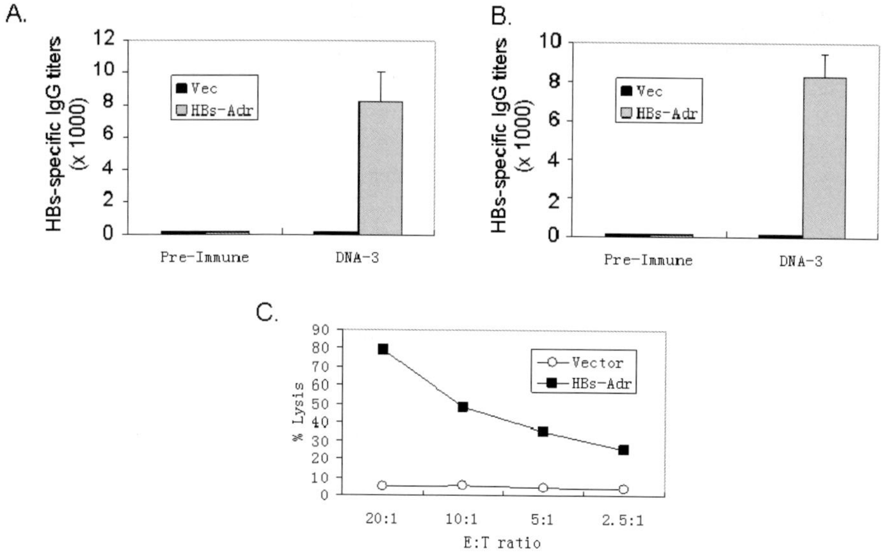


**Fig. 4.** HBs-specific antibody and CTL responses in C57/BL6 and BALB/C mice induced by the HBs-Adr DNA vaccine (HBs-Adr) or the empty vector (Vec). A, B: The antigen-specific IgG responses were measured by ELISA against HBs antigen. The peak level of HBc-specific IgG titers two weeks after the third DNA immunization by gene gun inoculation in C57BL/6 and BALB/C mice, respectively. The values represent mean IgG titers of 6 mice/group ± standard deviations. C: HBc-specific CTL responses four weeks after the third DNA immunization in BALB/C mice. Data are shown at different E:T ratios. The % lysis is expressed as the average of 6 mice/group.

70%–80% lysis at an effector:target ratio of 1:20. The vector control groups did not have any positive antibody or CTL responses.

### 2.2.3. *The effect of deglycosylation of HBs DNA vaccines*

Based on the results above that the HBs protein expressed DNA vaccine is N-glycosylated, we made HBs DNA vaccines with glycosylation mutations (HBs-dG) at various sites (Fig. 7A). First, we designed and constructed two versions of HBs-dG DNA vaccines using site-specific mutagenesis: HBs-dG1 and HBs-dG123, where N4 and N4/N114/N201 were removed, respectively (Fig. 5A). The site-specific mutations were

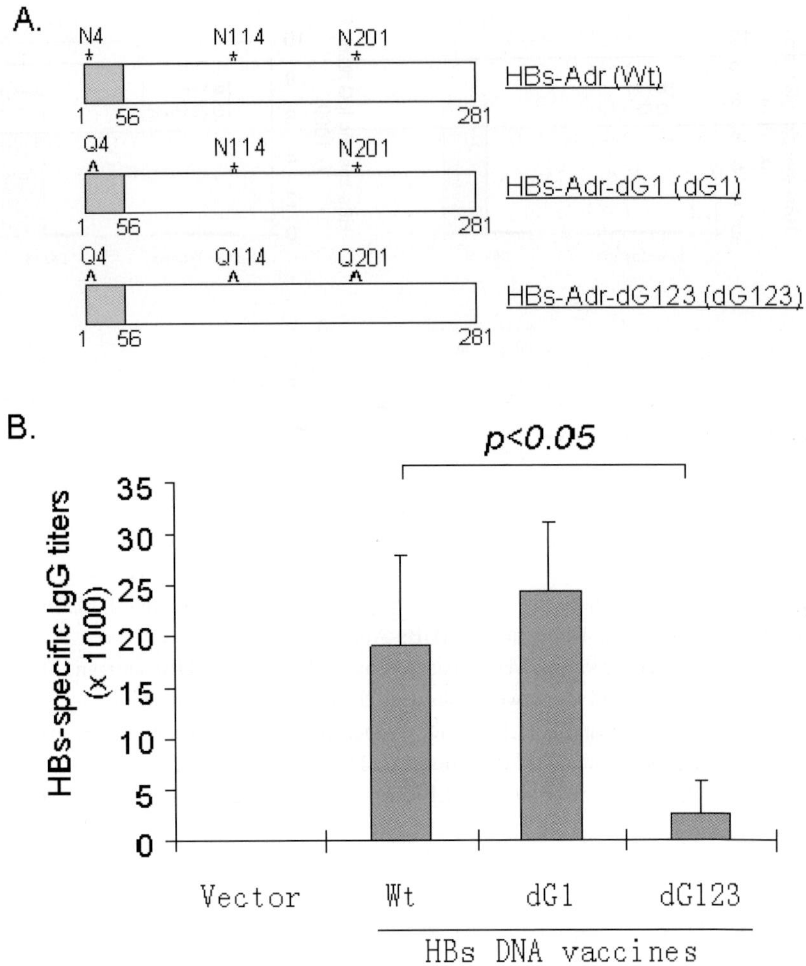


**Fig. 5.** A: Schematic diagram of HBs gene inserts in DNA vaccine constructs. HBs-Adr (Wt), HBs-Adr-dG1 (dG1) and HBs-Adr-dG123 (dG123) represent the wild type HBs, and two HBs with one or three mutations at the potential *N*-linked glycosylation (PNLG) sites of N4Q, N114Q and N201Q, respectively. * indicates the PNLG sites while ^ represents the site with an N-to-Q mutation. The gray and open boxes indicate the pre-S2 and S protein, respectively. The numbers indicate the amino acid positions. B: The HBs-specific antibody titers induced by various HBs-Adr DNA vaccines (Wt, dG1, dG123) with the empty vector as the negative control in BALB/C mice. The sera were collected two weeks after the third DNA immunization by gene gun inoculation. The values represent mean IgG titers of 5 mice/group ± standard deviations. The *P* value indicates the statistical significance between different DNA vaccine groups.

Asn to Gln (N–Q) mutations. All mutations were confirmed by sequencing analysis (data not shown). The expression of HBs-dG1 and HBs-dG123 proteins was confirmed by transient transfection of 293T cells and Western blot analysis. Interestingly, compared to the HBs wild type DNA vaccine, HBs-dG1 expressed a similar level of HBs antigen in both transfected cell lysate and supernatant, while HBs-dG123 expression was present only in the cell lysate but not supernatant, with a lower amount being expressed (data not shown). This finding indicates that removal of the N114 and N201 glycosylation sites may direct the HBs antigen to go through different post-translational processes. As expected, the molecular weight of the deglycosylated S protein was less than that of the wild type S protein, supporting the successful mutation of two PNLG sites.

In using the HBs-dG1 and -dG123 DNA vaccines to immunize BALB/C mice (5 mice/group) by gene gun inoculation, both HBs-dG DNA vaccines induced HBs-specific antibody responses. When compared to the wild type HBs DNA vaccine, HBs-dG1 induced a similar high level of HBs-specific antibody responses with end-point titration titers 1:15 000–1:30 000, while HBs-dG123 induced significantly lower titers (~1:2000; $P < 0.05$) (Fig. 5B). We further tested how dG1 and dG123 affected the T cell responses using an IFN-γ ELISPOT assay to measure the HBs-peptide-specific T cell responses in mouse splenocytes after the third DNA immunization. Both the HBs wild type and dG1 DNA vaccines generated a high level of specific T cell responses (400–500 IFN-γ spot-forming cells/million splenocytes). However, the HBs-dG123 DNA vaccine induced a much lower level of IFN-γ ELISPOT T cell responses as well ($P < 0.5$). The deglycosylation of N4 (dG1) did not affect HBs DNA vaccine expression and immunogenicity. Interestingly, when two additional glycosylation sites, N114 and N210, were mutated from dG1 (dG123), the expression and immunogenicity of the HBs-dG123 DNA vaccine were significantly depressed.

To further study the effect of removing only N114 and N210, we made one additional construct expressing mutants by site-specific N–Q mutagensis at N114 and N201, and an HBs-dG23 construct was made.[23] The expression of both wild type and deglycosylated S proteins was confirmed by detection of transient expression of both antigens in 293T

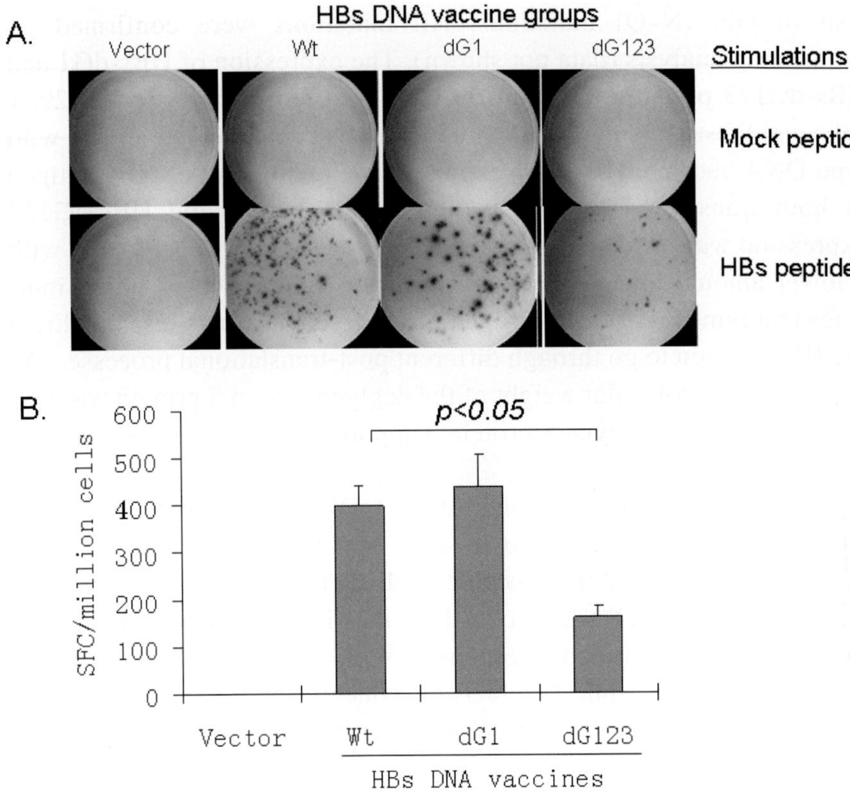


**Fig. 6.** ELISPOT analysis for IFN-γ secretion in mouse splenocytes immunized with HBs-Adr (Wt), HBs-Adr-dG1 (dG1) and HBs-Adr-dG123 (dG123) DNA vaccines. A: Actual sample wells with mock or HBs peptide stimulation. B: Frequency of HBs peptide-specific spots per million splenocytes from mice immunized with different HBs-Adr DNA vaccines (wt, dG1, DG123). Data represent the average of spot-forming cells (SFCs) per million splenocytes from 5 mice/group ± standard deviation. The spelnocytes were collected one week after the fourth DNA immunization. The *P* value indicates the statistical significance between different DNA vaccine groups.

cells and verified by Western blot.[23] High level antigen expression was achieved in both cell lysate and culture supernatant.

BALB/C mice were immunized with either the HBs-Wt (wild type) or HBs-dG23 DNA vaccines via a gene gun. S-antigen-specific antibody responses were detected in immunized animals by ELISA. High levels of anti-S-antibodies were elicited as measured by end titration titers of peak

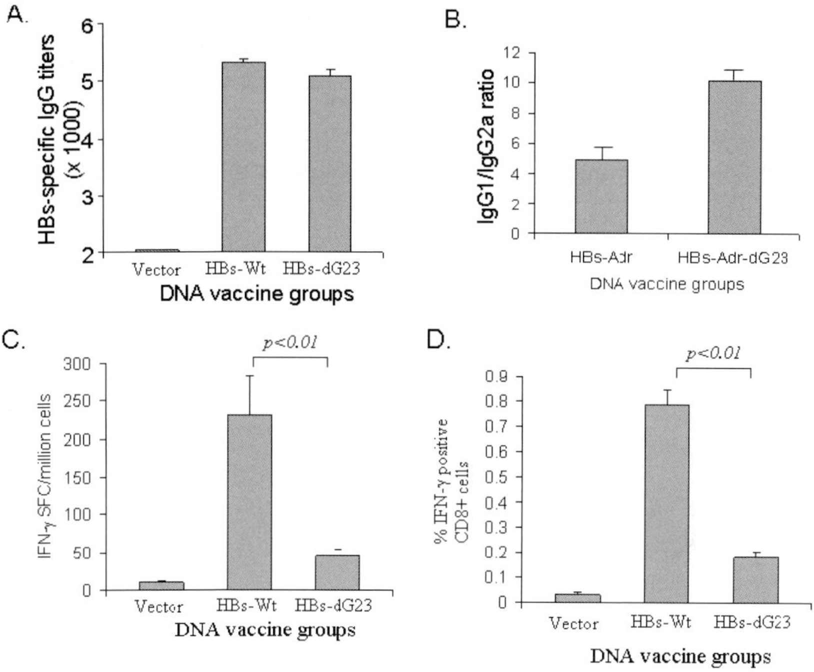


**Fig. 7.** HBs-specific antibody and T cell responses in BALB/C mice induced by HBs-Adr or HBs-Adr-dG23 DNA vaccines. The HBs-specific IgG responses one week after the fourth DNA immunization were measured by ELISA against HBs-Adr antigen (A and B). A: The titers represent the mean IgG titers of 5 mice/group ± standard deviations. B: The analysis of HBc-specific IgG isotypes (IgG1/IgG2a ratio). The values represent average IgG1/IgG2a ratios of 5 mice/group ± standard deviations. ELISPOT and intracellular cytokine staining (ICS) analyses were made for IFN-γ production in mouse splenocytes immunized with either HBs-Adr (Wt) or HBs-Adr-dG23 DNA vaccines (C and D). The spelnocytes were collected one week after the fourth DNA immunization. The *P* value indicates the statistical significance between different DNA vaccine groups. C: The frequency of HBs-peptide-specific spots per million splenocytes in mice immunized with different HBs-Adr DNA vaccines. Data represent the average of spot-forming cells (SFCs) per million splenocytes from 5 mice/group ± standard deviation. D: Percentage of IFN-γ-producing cells specific for HBs peptide. Data represent the average percentage of IFN-γ-producing cells from 5 mice/group ± standard deviation.

level animal sera (Fig. 7A). Overall, very similar antibody responses were generated by both the HBs-Wt and HBs-dG23 groups (Fig. 7A).[23] Further analysis of the subtypes of IgG responses showed a predominant Th2 type antibody response to HBs, as measured by the IgG1/IgG2a ratio in both groups of immunized mouse sera (Fig. 7B). This is inconsistent with a previous finding that gene gun-mediated DNA immunization mainly elicits a Th2 type immune response.[6,10,13,14] The HBs-dG23 DNA was able to induce an even higher ratio of IgG1/IgG2a (Fig. 9B), suggesting an even stronger Th2 response when compared to the wild type DNA vaccine.[23]

Interestingly, the ability of HBs-dG23 DNA vaccines to elicit S-specific cell-mediated immune (CMI) responses was different from the wild type HBs DNA vaccine. Mice immunized with the HBs-dG123 DNA vaccine had significantly lower CMI responses when compared to those that received the wild type HBs DNA vaccine. IFN-γ responses, as measured by ELISPOT, consisted of only 25% of the responses elicited by the wild type HBs-Adr DNA vaccine (Fig. 7C).[23] Similarly, IFN-γ expressing CD8+ T cells in the HBs-Adr-dG23 DNA vaccine group were less than 1/3 of that elicited by the wild type HBs-Adr DNA vaccine, as measured by flow cytometry (Fig. 7D).[23]

## 2.3. *Immunogenicity of HBc and HBs DNA vaccines in nonhuman primates*

### 2.3.1. *Induction of HBc-specific immune responses by the HBc DNA vaccine in rhesus macaques*

Two groups of macaques (2 per group) were immunized with the DNA vector pJW4303 or the HBc DNA vaccine, respectively, by IM inoculation every two months. The HBc-specific antibody kinetics showed that the antibody responses appeared after the first immunization and reached the peak level after the fourth DNA immunization. The animals that received the HBc DNA vaccine produced a high level of HBc-specific antibody responses (titers 1:50 000–100 000) when compared to the animals receiving the DNA vector control (Fig. 8A) as measured two months after the last DNA immunization. The IgG isotype analysis showed that HBc DNA immunzation by IM in macaques induced predominantly Th1 type antibody responses with the IgG1/IgG2a ratio <1.0 (Fig. 8B).

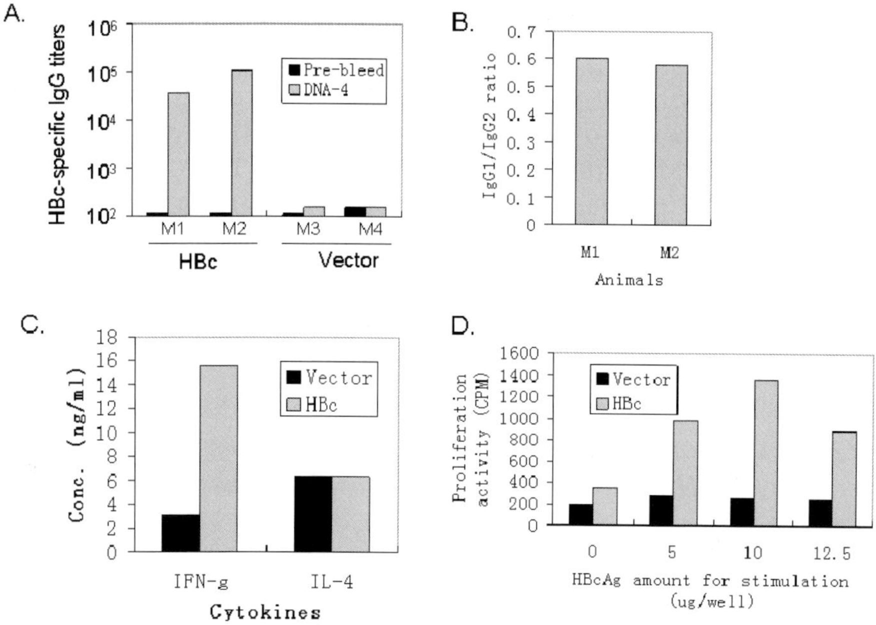


**Fig. 8.** HBc-specific antibody and T cell responses in macaques induced by intramuscular injection of the HBc DNA vaccine (HBc) or the empty vector. The HBc-specific IgG responses were measured by ELISA (A and B). A: HBc-specific IgG responses in individual animal sera collected prior to immunization (prebleed) and four weeks after the fourth DNA immunization (DNA-4). B: The analysis of HBc-specific IgG isotypes (IgG1/IgG2a ratio). Immune sera were collected four weeks after the fourth DNA immunization (DNA-4). Cytokine production and T cell proliferation activity analyses were made for HBc-specific T cell responses in PBMCs two months after the fourth DNA immunization, respectively (C and D). C: IFN-γ and IL-2 levels in culture supernatant after HBcAg stimulation. D: T cell proliferation activity after stimulation with different amounts of HbcAg, as indicated. Data represent the average of two monkeys in each group.

The HBc-specific T cell responses were examined in PBMCs isolated two months after the fourth DNA immunization. Following stimulation of PBMCs with recombinant HBcAg, IFN-γ and IL-4 levels in culture supernatant were detected. The PBMC culture of animals receiving HBc DNA vaccines produced a high level of IFN-γ when compared to the animals in the vector control group (Fig. 8C). There was no difference in the

IL-4 level in the PBMC culture between the HBc- and vector-immunized groups (Fig. 8C).

HbcAg-specific proliferation activities were also measured in PBMCs of macaques with or without recombinant HBcAg stimulation. Clearly, the PBMCs from animals receiving HBc DNA vaccines had a high level of proliferation activities after the HBcAg stimulation when compared to the PBMCs from the vector group or to no stimulation. The above results indicated that the HBc DNA vaccine was able to induce both antibody and T cell responses against HBc in rhesus macaques.

### 2.3.2. *Induction of HBs-specific immune responses by the HBs DNA vaccine in rhesus macaques*

Two groups of macaques (2 per group) were immunized with the DNA vector pSW3891 or the HBs DNA vaccine, respectively, by IM inoculation every eight weeks for four immunizations, followed by the fifth immunization at week 48. HBs-specific antibody kinetics showed that antibody responses appeared after the first immunization and reached the peak level after the fifth DNA immunization. The animals receiving HBs DNA vaccines produced a high level of HBs-specific antibody responses (titers 1:10 000) compared to the animals that receiving the DNA vector control (Fig. 9A) as measured two months after the fifth DNA immunization; the HBs-specific responses persisted at a high level till week 125. The HBs-specific T cell responses were examined in PBMCs isolated four weeks after the fifth DNA immunization using HBsAg-specific proliferation assay and IFN-γ ELISPOT. Clearly, the PBMCs from animals receiving the HBs DNA vaccine had a high level of proliferation activities after the HBsAg stimulation when compared to the PBMCs from the vector group or to no stimulation. The animals receiving the HBs DNA vaccine also had detectable IFN-γ spot-forming cells (40–100 spots/million PBMCs) after the HBsAg stimulation when compared to the vector control group (< 5 spots/million PBMCs). These results demonstrated that the HBs DNA vaccines effectively induced both antibody and T cell responses to HBs in rhesus macaques.

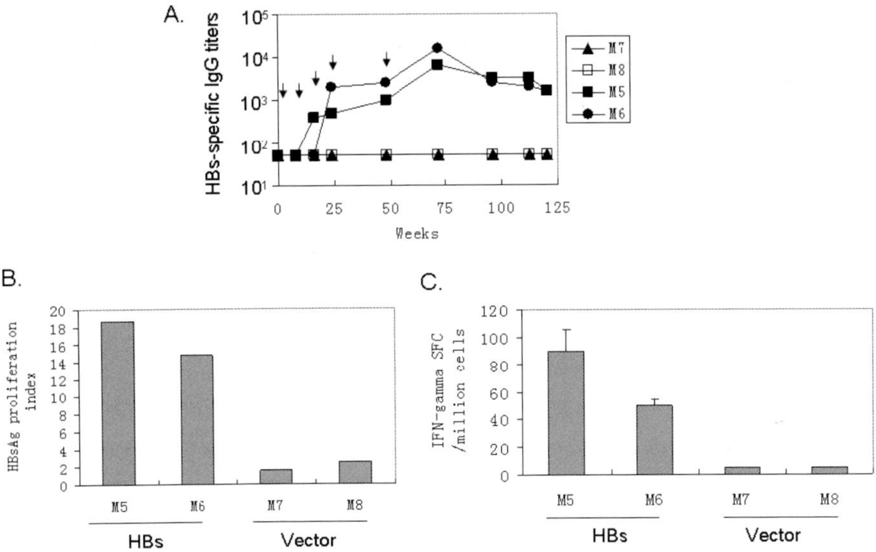


**Fig. 9.** HBs-specific antibody and T cell responses in macaques induced by intramuscular injection of the HBs DNA vaccine (HBs) or the empty vector. A: Temporal HBs-specific IgG responses were measured by ELISA. Each curve represents the HBs-specific antibody response for each animal. Animals M5 and M6 received the HBs DNA vaccine, while M7 and M8 received the empty vector. The arrows indicate the time of DNA immunizations. T cell proliferation activity and IFN-γ ELISPOT analyses were made for HBs-specific T cell responses in PBMCs of individual animals collected four weeks after the fifth DNA immunization, respectively (B and C). B: T cell proliferation activity of PBMCs after stimulation with HBsAg. C: The frequency of HBsAg-specific spots per million PBMCs in HBs-Adr DNA vaccine- or vector-immunized macaques.

## 3. Discussion

In this report, results from our HBV DNA vaccine studies conducted in the last ten years are summarized. Our data demonstrate that hepatitis B (HBV) DNA vaccines induced high levels of HBcAg- and HBsAg-specific antibody and T cell responses in both mice and nonhuman primates. Furthermore, HBV-specific immune responses can be enhanced by using either a more effective DNA vaccine delivery system or an adjuvant.

The modification/mutation of Asn (*N*)-linked glycosylation in S protein affects the HBsAg-specific immune responses. These findings are important for the ultimate utility of HBV DNA vaccines as part of the novel therapeutic strategy for chronically infected HBV populations.

The HBV DNA vaccine was among the first to show the immunogenicity of DNA vaccines. In addition, the HBV DNA vaccine expressing the S antigen was the first-ever DNA vaccine to elicit protective level immune responses in humans.[17] In this landmark study, an HBV S DNA vaccine elicited balanced cell-mediated and antibody responses to the HBV S antigen. This DNA plasmid included an HBsAg gene insert containing a portion of the pre-S2 sequence from the strain *adw.* After only two immunizations, 9 out of 12 volunteers seroconverted and 3 volunteers developed seroprotective levels of antibody (> 10 mIU/ml), and after the third immunization all 12 of the volunteers had developed seroprotective levels. By the third immunization, all 12 of the vaccinees displayed significant numbers of HBsAg-specific IFN-$\gamma$ spot-forming cells. Furthermore, CD+8 T cell responses were detected in all 12 volunteers by flow-cytometric analysis. Subsequent studies demonstrated that the same highly immunogenic HBV S DNA vaccine was able to elicit anti-S-antibody responses in volunteers who were previously unresponsive to a traditional HBV vaccine based on recombinant S protein.[16]

HBV DNA vaccine studies in China started in the mid-1990s and have continued to be an active area of research. Results described in the current report appeared earlier than the above reported human studies but did not have a chance to move into human studies given the limited regulatory review of DNA vaccines in China. However, our results have yielded a number of interesting observations that will be useful for the development of the next generation of improved HBV DNA vaccines. Most importantly, it is clear now that the delivery method of the DNA vaccine plays a critical role in the induction of detectable immune responses in humans. With the recent emerging interest in electroporation as a more effective DNA vaccine delivery method than the traditional needle injection approach, we can anticipate renewed interest in using the HBV DNA vaccine as a possible therapeutic approach to chronic HBV infection. Data presented in this report can serve as useful reference for the progress of

HBV DNA vaccines in China up to this point. Further optimized HBV DNA vaccines can also incorporate information gained from this report to develop next generation immunotherapies to control HBV, a major human infectious disease in China.

## References

1. Davis HL *et al.*, *Hum Gene Ther*, 1995, **6**: 1447–1456.
2. Fissolo N *et al.*, *Eur J Immunol*, 2005, **35**: 117–127.
3. Glebe D *et al.*, *J Virol*, 2003, **77**: 9511–9521.
4. Hanke T, *Eur J Immunol*, 2006, **36**: 806–809.
5. Huang CF *et al.*, *Cell Mol Immunol*, 2006, **3**: 97–106.
6. Koziel MJ, *Antivir Ther*, 1998, **3**: 13–24.
7. Kuhrober A *et al.*, *Int Immunol*, 1997, **9**: 1203–1212.
8. Lee HG *et al.*, *Virus Res*, 1997, **50**: 185–194.
9. McGovern BH, *Antivir Ther 12 (Suppl)*, 2007, **3**: H3–H13.
10. Mehta A *et al.*, *Adv Exp Med Biol*, 1998, **435**: 195–205.
11. Milich DR *et al.*, *J Virol*, 1997, **71**: 2192–201.
12. Oka Y *et al.*, *Immunology*, 2001, **103**: 90–97.
13. Pichoud C *et al.*, *J Hepatol*, 2000, **32**: 307–316.
14. Pontisso P *et al.*, *J Hepatol*, 1991, **12**: 203–206.
15. Rico MA *et al.*, *Hepatology*, 2001, **33**: 295–300.
16. Rottinghaus ST *et al.*, *Vaccine*, 2003, **21**: 4604–4608.
17. Roy MJ *et al.*, *Vaccine*, 2000, **19**: 764–778.
18. Shimada N *et al.*, *J Clin Immunol*, 2003, **23**: 223–232.
19. Townsend K *et al.*, *J Virol*, 1997, **71**: 3365–3374.
20. Tsai SL *et al.*, *J Gastroenterol Hepatol*, 1997, **12**: S227–S35.
21. Wang S *et al.*, *Vaccine*, 2008, **26**: 2100–2110.
22. He XW *et al.*, *Vaccine*, 2005, **23**: 1649–1656.
23. Xing Y, *et al.*, *Vaccine*, 2008, **26**: 5145–5152.

# List of Contributors

**Xia Cai**
Drug Discovery and Design Center
State Key Laboratory of Drug Research
Shanghai Institute of Materia Medica
Chinese Academy of Sciences
Shanghai 201203, China
*Chapter 1.5*

**Yan-Shan Cai**
Guangzhou Center for Disease Control and Prevention
Guangzhou 510080, China
*Chapter 3.6*

**Jian Cao**
College of Bioengineering
Henan University of Technology
Zhengzhou 450052, China
*Chapter 1.3*

**Chao Chen**
Department of Neonatology
Children's Hospital of Fudan University
Shanghai 201102, China
*Chapter 3.4*

**Zhe Chen**
Key Laboratory of Medical Molecular Virology
Shanghai Medical College
Institutes of Biomedical Sciences
Fudan University
Shanghai 200032, China
*Chapter 2.2*

**Jing-Xian Chen**
Department of Microbiology
Anhui Medical University
The Key Laboratory of Gene Resource Utilization for Genetic Diseases
Ministry of Education and Anhui Province
Hefei 230032, China
*Chapter 3.7*

**Ming-Liang Chen**
Key Laboratory of Medical Molecular Virology
Shanghai Medical College
Institutes of Biomedical Sciences
Fudan University
Shanghai 200032, China
*Chapter 1.5*

**Li-Juan Chunyu**
State Key Laboratory of Genetic Engineering
Institute of Genetics
School of Life Sciences
Fudan University
Shanghai 200433, China
*Chapter 1.3*

**Biao Di**
Guangzhou Center for Disease Control and Prevention
Guangzhou 510080, China
*Chapter 3.6*

**Xing Ding**
Department of Pulmonary Medicine
Shanghai First People's Hospital
Jiao Tong University
Shanghai 200080, China
*Chapter 4.1*

**Sheng-Fu Dong**
Key Laboratory of Medical Molecular Virology
Shanghai Medical College
Institutes of Biomedical Sciences
Fudan University
Shanghai 200032, China
*Chapter 1.5, Chapter 2.3*

**Tong-Hai Dou**
Institute of Genetics
Fudan University
Shanghai 200433, China
*Chapter 1.6*

**Kun Du**
Department of Immunology
Xiangya Medical School of Central South University
Changsha 410008, China
*Chapter 3.5*

**Xiao-Hong Du**
Department of Dermatology
Shenzhen People's Hospital
Second Clinical Medicine College of Jinan University
Shenzhen 518020, China
*Chapter 5.1*

**Hong-Jie Fan**
Key Laboratory of Animal Disease Diagnostic and Immunology
Ministry of Agriculture
Nanjing Agriculture University
Nanjing 210095, China
*Chapter 1.4*

**Cai-Yun Fang**
Institutes of Biomedical Sciences
Fudan University
Shanghai 200032, China
*Chapter 2.1*

**Si-Ning Fang**
Department of Dermatology
Shenzhen People's Hospital
Second Clinical Medicine College of Jinan University
Shenzhen 518020, China
*Chapter 5.1*

**Long-Ting Fu**
Shanghai Public Health Clinical Center affiliated to Fudan University
Shanghai 201508, China
*Chapter 3.1*

**Kai Gao**
Guangzhou Center for Disease Control and Prevention
Guangzhou 510080, China
*Chapter 3.8*

**Qi-Ming Gong**
Department of Infectious Diseases
Ruijin Hospital
Shanghai Jiao Tong University
School of Medcine
Shanghai 200025, China
*Chapter 4.4*

**Ming Gu**
Shanghai Entry and Export Inspection and Quarantine Bureau
Shanghai 200135, China
*Chapter 1.6*

**Guo-Hao Gu**
Department of Medical Laboratory Science
The First Affiliated Hospital of Soochow University
Suzhou 215006
China
*Chapter 3.9*

**Jian-Ping Gu**
Deqing County Center for Disease Control and Prevention
Deqing 313200, China
*Chapter 2.5*

**Shi-Min Gu**
Shanghai Public Health Clinical Center affiliated to Fudan University
Shanghai 201508, China
*Chapter 3.1*

**Yu-Dong Gu**
Department of Hand Surgery
Huashan Hospital
Fudan University
Shanghai 200031, China
*Chapter 4.6*

**Xiao-Kui Guo**
Department of Medical Microbiology and Parasitology
Institute of Medical Sciences
Shanghai Jiao Tong University
School of Medicine
Shanghai 200025, China
*Chapter 1.1, Chapter 1.2*

**Cong Han**
Key Laboratory of Medical Molecular Virology
Shanghai Medical College
Institutes of Biomedical Sciences
Fudan University
Shanghai 200032, China
*Chapter 1.5*

**Xiao-Ling Han**
Key Laboratory of Medical Molecular Virology
Shanghai Medical College
Institutes of Biomedical Sciences
Fudan University
Shanghai 200032, China
*Chapter 2.3*

**Zhi-Gang Han**
Guangzhou Center for Disease Control and Prevention
Guangzhou 510080, China
*Chapter 3.8*

**Bao-Yu Hu**
Department of Medical Microbiology and Parasitology
Institute of Medical Sciences
Shanghai Jiao Tong University
School of Medicine
Shanghai 200025, China
*Chapter 1.1*

**Yun-Wen Hu**
Key Laboratory of Medical Molecular Virology
Shanghai Medical College
Institutes of Biomedical Sciences
Fudan University
Shanghai 200032, China
*Chapter 5.3*

**Rui Huang**
Department of Microbiology
Medical College of Soochow University
Suzhou 215123, China
*Chapter 3.9*

**Yu-Wen Huang**
Shanghai Blood Center
Shanghai 200051, China
*Chapter 2.3*

**Zu-Hu Huang**
Jiangsu Province Infectious Diseases Key Laboratory
The First Affiliated Hospital
Nanjing Medical University
Nanjing 210029, China
*Chapter 5.4*

**Zhi Huo**
Department of Immunology
Xiangya Medical School of Central South University
Changsha 410008, China
*Chapter 3.5*

**Qing-Wu Jiang**
Key Laboratory of Public Health Safety
Ministry of Education
Department of Epidemiology
School of Public Health
Fudan University
Shanghai 200032, China
*Chapter 2.5*

**Wei-Lun Jiang**
Shanghai Public Health Clinical Center affiliated to Fudan University
Shanghai 201508, China
*Chapter 3.1*

**Xiu-Gao Jiang**
National Institute for Communicable Disease Control and Prevention
Chinese Center for Disease Control and Prevention
Beijing 102206, China
*Chapter 1.2*

**Wei Jiao**
Key Laboratory of Public Health Safety
Ministry of Education
Department of Epidemiology
School of Public Health
Fudan University
Shanghai 200032, China
*Chapter 2.5*

**Jia-Lin Jin**
Huashan Hospital
Fudan University
Shanghai 200040, China
*Chapter 3.2*

**Rui-Liang Jin**
State Key Laboratory of Genetic Engineering
Institute of Genetics
School of Life Sciences
Fudan University
Shanghai 200433, China
*Chapter 1.3*

**Jin-Shan Ke**
Department of Microbiology
Second Military Medical University
Shanghai 200433, China
*Chapter 5.2*

**Jian-Qiang Lei**
Huashan Hospital
Fudan University
Shanghai 200040, China
*Chapter 3.2*

**Jun Li**
China–US Vaccine Research Center
The First Affiliated Hospital
Nanjing Medical University
Nanjing 210029, China
*Chapter 5.4*

**Yan Li**
Guangzhou Center for Disease Control and Prevention
Guangzhou 510080, China
*Chapter 3.8*

**Fang-Juan Li**
Department of Dermatology
Shenzhen People's Hospital
Second Clinical Medicine College of Jinan University
Shenzhen 518020, China
*Chapter 5.1*

**Qi-Han Li**
Institute of Medical Biology
Chinese Academy of Medical Sciences
Peking Union Medical College
Kunming 650118, China
*Chapter 2.4*

**Xiu-Wen Li**
National Institute for Communicable Disease Control and Prevention
Chinese Center for Disease Control and Prevention
Beijing 102206, China
*Chapter 1.2*

**Yi-Xue Li**
Shanghai Center for Bioinformation Technology
Shanghai 200235, China
*Chapter 2.2*

**Yan Liang**
Institute of Medical Biology
Chinese Academy of Medical Sciences
Peking Union Medical College
Kunming 650118, China
*Chapter 2.4*

**Cai-Yun Liang**
Guangzhou Center for Disease Control and Prevention
Guangzhou 510080, China
*Chapter 3.8*

**Wei Lin**
Licheng Center for Disease Control and Prevention
Jinan 250100, China
*Chapter 1.1*

**Bing Ling**
Shanghai Blood Center
Shanghai 200051, China
*Chapter 2.3*

**Min Liu**
Department of Microbiology
Second Military Medical University
Shanghai 200433, China
*Chapter 5.2*

**Qian Liu**
Department of Microbiology
Anhui Medical University
The Key Laboratory of Gene Resource Utilization for Genetic Diseases
Ministry of Education and Anhui Province
Hefei 230032, China
*Chapter 3.7*

**Yang Liu**
Guangzhou Center for Disease Control and Prevention
Guangzhou 510080, China
*Chapter 3.6*

**Long-Ding Liu**
Institute of Medical Biology
Chinese Academy of Medical Sciences
Peking Union Medical College
Kunming 650118, China
*Chapter 2.4*

**Rui-Qing Liu**
State Key Laboratory of Genetic Engineering
Institute of Genetics
School of Life Sciences
Fudan University
Shanghai 200433, China
*Chapter 1.3*

**Yu-Fei Liu**
Guangzhou Center for Disease Control and Prevention
Guangzhou 510080, China
*Chapter 3.6*

**Jian-Er Long**
Key Laboratory of Medical Molecular Virology
Shanghai Medical College
Institutes of Biomedical Sciences
Fudan University
Shanghai 200032, China
*Chapter 2.3*

**Ping Lu**
Shanghai Blood Center
Shanghai 200051, China
*Chapter 2.3*

**Shan Lu**
Laboratory of Nucleic Acid Vaccines
Department of Medicine
University of Massachusetts Medical School
Worcester, MA 01605, USA
*Chapter 5.4*

**Cheng-Ping Lu**
Key Laboratory of Animal Disease Diagnostic and Immunology
Ministry of Agriculture
Nanjing Agriculture University
Nanjing 210095, China
*Chapter 1.4*

**En-Jie Lu**
Guangzhou Center for Disease Control and Prevention
Guangzhou 510080, China
*Chapter 3.6*

**Yi-Han Lu**
Key Laboratory of Public Health Safety
Ministry of Education
Department of Epidemiology
School of Public Health
Fudan University
Shanghai 200032, China
*Chapter 2.5*

**Zhi-Meng Lu**
Department of Infectious Diseases
Ruijin Hospital
Shanghai Jiao Tong University
School of Medcine
Shanghai 200025, China
*Chapter 4.4*

**Shao-Hui Ma**
Institute of Medical Biology
Chinese Academy of Medical Sciences
Peking Union Medical College
Kunming 650118, China
*Chapter 2.4*

**Shu-Hui Ma**
Department of Immunology
Xiangya Medical School of Central South University
Changsha 410008, China
*Chapter 3.5*

**Zhang-Mei Ma**
Key Laboratory of Medical Molecular Virology
Shanghai Medical College
Institutes of Biomedical Sciences
Fudan University
Shanghai 200032, China
*Chapter 2.1*

**Bing Mei**
Department of Immunology
Xiangya Medical School of Central South University
Changsha 410008, China
*Chapter 3.5*

**Cheng-Yan Meng**
Huashan Hospital
Fudan University
Shanghai 200040, China
*Chapter 3.2*

**Xiao-Hui Miao**
Department of Infectious Diseases
Changzheng Hospital
Second Military Medical University
Shanghai 200040, China
*Chapter 4.6*

**Qin Mo**
Shanghai Blood Center
Shanghai 200051, China
*Chapter 2.3*

**Yi-Xin Nie**
National Institute for Communicable Disease Control and Prevention
Chinese Center for Disease Control and Prevention
Beijing 102206, China
*Chapter 1.2*

**Wei Pan**
Department of Microbiology
Second Military Medical University
Shanghai 200433, China
*Chapter 5.2*

**Kai-Cheng Qian**
Shanghai Blood Center
Shanghai 200051, China
*Chapter 2.3*

**Hui Qi**
Clinical Medical Research Center
Shenzhen People's Hospital
Second Clinical Medicine College of Jinan University
Shenzhen 518020, China
*Chapter 5.1*

**Zhong-Tian Qi**
Department of Microbiology
Second Military Medical University
Shanghai 200433, China
*Chapter 5.2*

**Peng-Zhe Qin**
Guangzhou Center for Disease Control and Prevention
Guangzhou 510080, China
*Chapter 3.6*

**Zhi-Qiang Qin**
Key Laboratory of Medical Molecular Virology
Shanghai Medical College
Institutes of Biomedical Sciences
Fudan University
Shanghai 200032, China
*Chapter 1.5*

**De-Qi Qiu**
Department of Infectious Diseases
Ruijin Hospital
Shanghai Jiao Tong University
School of Medicine
Shanghai 200025, China
*Chapter 4.4*

**Di Qu**
Key Laboratory of Medical Molecular Virology
Shanghai Medical College
Institutes of Biomedical Sciences
Fudan University
Shanghai 200032, China
*Chapter 1.5, Chapter 2.3, Chapter 3.1*

**Jun Ren**
Key Laboratory of Medical Molecular Virology
Shanghai Medical College
Institutes of Biomedical Sciences
Fudan University
Shanghai 200032, China
*Chapter 2.1*

**Fei-Yi Ruan**
Huashan Hospital
Fudan University
Shanghai 200040, China
*Chapter 3.2*

**Jun-Jie Shao**
Shanghai Municipal Center for Disease Control and Prevention
Shanghai 200336, China
*Chapter 4.2*

**Fang Shen**
Shanghai Public Health Clinical Center affiliated to Fudan University
Shanghai 201508, China
*Chapter 3.1*

**Jun Shen**
Children's Hospital of Fudan University
Shanghai 201102, China
*Chapter 4.5*

**Xu Shen**
Drug Discovery and Design Center
State Key Laboratory of Drug Research
Shanghai Institute of Materia Medica
Chinese Academy of Sciences
Shanghai 201203, China
*Chapter 1.5*

**Ji-Chuan Shen**
Guangzhou Center for Disease Control and Prevention
Guangzhou 510080, China
*Chapter 3.6*

**Dong-Mei Shi**
Department of Infectious Diseases
Ruijin Hospital
Shanghai Jiao Tong University
School of Medicine
Shanghai 200025, China
*Chapter 4.3*

**Li-Yun Sun**
Key Laboratory of Animal Disease Diagnostic and Immunology
Ministry of Agriculture
Nanjing Agriculture University
Nanjing 210095, China
*Chapter 1.4*

**Li-Zhi Tan**
Institute of Pathogenic Biology
Nan Hua University
Hengyang
Hengyang 421001, China
*Chapter 1.1*

**Zheng Teng**
Shanghai Municipal Center for Disease Control and Prevention
Shanghai 200336, China
*Chapter 4.2*

**Qiu-Sheng Tong**
Department of Dermatology
Shenzhen People's Hospital
Second Clinical Medicine College of Jinan University
Shenzhen 518020, China
*Chapter 5.1*

**Jian Wang**
Key Laboratory of Medical Molecular Virology
Shanghai Medical College
Institutes of Biomedical Sciences
Fudan University
Shanghai 200032, China
*Chapter 5.3*

**Ming Wang**
Guangzhou Center for Disease Control and Prevention
Guangzhou 510080, China
*Chapter 3.6*

**Qi Wang**
Drug Discovery and Design Center
State Key Laboratory of Drug Research
Shanghai Institute of Materia Medica
Chinese Academy of Sciences
Shanghai 201203, China
*Chapter 1.5*

**Chuan-Qin Wang**
Children's Hospital of Fudan University
Shanghai 201102, China
*Chapter 4.5*

**Fa-Di Wang**
Deqing County Center for Disease Control and Prevention
Deqing 313200, China
*Chapter 2.5*

**Fu-Yan Wang**
Department of Immunology
Xiangya Medical School of Central South University
Changsha 410008, China
*Chapter 3.5*

**Hong-Hai Wang**
State Key Laboratory of Genetic Engineering
Institute of Genetics
School of Life Sciences
Fudan University
Shanghai 200433, China
*Chapter 1.3*

**Hong-Jun Wang**
State Key Laboratory of Genetic Engineering
Institute of Genetics
School of Life Sciences
Fudan University
Shanghai 200433, China
*Chapter 1.3*

**Jing-Jing Wang**
Institute of Medical Biology
Chinese Academy of Medical Sciences
Peking Union Medical College
Kunming 650118, China
*Chapter 2.4*

**Jun-Xue Wang**
Department of Infectious Diseases
Changzheng Hospital
Second Military Medical University
Shanghai 200040, China
*Chapter 4.6*

**Li-Chun Wang**
Institute of Medical Biology
Chinese Academy of Medical Sciences
Peking Union Medical College
Kunming 650118, China
*Chapter 2.4*

**Ming-Li Wang**
Department of Microbiology
Anhui Medical University
The Key Laboratory of Gene Resource Utilization for Genetic Diseases
Ministry of Education and Anhui Province
Hefei 230032, China
*Chapter 3.7*

**Shi-Xia Wang**
China–US Vaccine Research Center
The First Affiliated Hospital
Nanjing Medical University
Nanjing 210029, China;
*Chapter 5.4*

**Wen-Yi Wang**
Key Laboratory of Medical Molecular Virology
Shanghai Medical College
Institutes of Biomedical Sciences
Fudan University
Shanghai 200032, China
*Chapter 2.3*

**Xiao-Hong Wang**
Children's Hospital of Fudan University
Shanghai 201102, China
*Chapter 4.5*

**Xin-Gen Wang**
Clinical Medical Research Center
Shenzhen People's Hospital
Second Clinical Medicine College of Jinan University
Shenzhen 518020, China
*Chapter 5.1*

**Xue-Cai Wang**
Deqing County Center for Disease Control and Prevention
Deqing 313200, China
*Chapter 2.5*

**Yu-Lin Wang**
Guangzhou Center for Disease Control and Prevention
Guangzhou 510080, China
*Chapter 3.6*

**Kang-Sheng Wen**
Shanghai Municipal Center for Disease Control and Prevention
Shanghai 200336, China
*Chapter 4.2*

**Yu-Mei Wen**
Key Laboratory of Medical Molecular Virology
Shanghai Medical College
Institutes of Biomedical Sciences
Fudan University
Shanghai 200032, China
*Chapter 2.1, Chapter 2.2*

**Shu-Yan Wu**
Department of Microbiology
Medical College of Soochow University
Suzhou 215123, China
*Chapter 3.9*

**Zhong-Liang Wu**
Biomerieux Industry China
Shanghai 200001, China
*Chapter 1.6*

**Qing Xie**
Department of Infectious Diseases
Ruijin Hospital
Shanghai Jiao Tong University
School of Medicine
Shanghai 200025, China
*Chapter 4.3*

**Yi Xie**
Huashan Hospital
Fudan University
Shanghai 200025, China
*Chapter 5.3*

**Yi-Ping Xing**
Jiangsu Province Infectious Diseases Key Laboratory
The First Affiliated Hospital
Nanjing Medical University
Nanjing 210029, China
*Chapter 5.4*

**Ai-Qin Xu**
Deqing County Center for Disease Control and Prevention
Deqing 313200, China
*Chapter 2.5*

**Hui-Fang Xu**
Guangzhou Center for Disease Control and Prevention
Guangzhou 510080, China
*Chapter 3.8*

**Sheng-Feng Xu**
State Key Laboratory of Genetic Engineering
Institute of Genetics
School of Life Sciences
Fudan University
Shanghai 200433, China
*Chapter 1.3*

**Wen-Sheng Xu**
Department of Infectious Diseases
Changzheng Hospital
Second Military Medical University
Shanghai 200040, China
*Chapter 4.6*

**Yun-Min Xu**
State Key Laboratory of Genetic Engineering
Institute of Genetics
School of Life Sciences
Fudan University
Shanghai 200433, China
*Chapter 1.3*

**Yang Yang**
Department of Medical Microbiology and Parasitology
Institute of Medical Sciences
Shanghai Jiao Tong University
School of Medicine
Shanghai 200025, China
*Chapter 1.1*

**Hong-Liang Yang**
Institute of Plant Physiology and Ecology
Shanghai Institutes for Biological Sciences
Chinese Academy of Sciences
Shanghai 200032, China
*Chapter 1.1*

**Jie-Lin Yang**
Shanghai Entry and Export Inspection and Quarantine Bureau
Shanghai 200135, China
*Chapter 1.6*

**Peng-Yuan Yang**
Institutes of Biomedical Sciences
Fudan University
Shanghai 200032, China
*Chapter 2.1*

**Wen-Jun Yang**
Shiping County Center for Disease Control and Prevention
Shiping 662200, China
*Chapter 2.4*

**Zi-Jiang Yang**
Institute of Medical Biology
Chinese Academy of Medical Sciences
Peking Union Medical College
Kunming 650118, China
*Chapter 2.4*

**Wen-Wu Yin**
Chinese Center for Disease Control and Prevention
Beijing 100050, China
*Chapter 1.2*

**Ping Yu**
Department of Immunology
Xiangya Medical School of Central South University
Changsha 410008, China
*Chapter 3.5*

**Zheng-Hong Yuan**
Key Laboratory of Medical Molecular Virology
Shanghai Medical College
Institutes of Biomedical Sciences
Fudan University
Shanghai 200032, China
*Chapter 3.1, Chapter 5.3,*

**Jian Zhang**
Drug Discovery and Design Center
State Key Laboratory of Drug Research
Shanghai Institute of Materia Medica
Chinese Academy of Sciences
Shanghai 201203, China
*Chapter 1.5*

**Min Zhang**
Department of Pulmonary Medicine
Shanghai First People's Hospital
Jiao Tong University
Shanghai 200080, China
*Chapter 4.1*

**Rong Zhang**
Department of Neonatology
Children's Hospital of Fudan University
Shanghai 201102, China
*Chapter 3.4*

**Xi Zhang**
Shanghai Municipal Center for Disease Control and Prevention
Shanghai 200336, China
*Chapter 4.2*

**Ji-Ming Zhang**
Huashan Hospital
Fudan University
Shanghai 200025, China
*Chapter 5.3*

**Ren-Fang Zhang**
Shanghai Public Health Clinical Center affiliated to Fudan University
Shanghai 201508, China
*Chapter 3.1*

**Rong-Song Zhang**
Honghe Prefecture Center for Disease Control and Prevention
Mengzi 661100, China
*Chapter 2.4*

**Wen-Hong Zhang**
Huashan Hospital
Fudan University
Shanghai 200040, China
*Chapter 3.2*

**Xiang-Yan Zhang**
Department of Medical Microbiology and Parasitology
Institute of Medical Sciences
Shanghai Jiao Tong University
School of Medicine
Shanghai 200025, China
*Chapter 1.2*

**Xin-Xin Zhang**
Department of Infectious Diseases
Ruijin Hospital
Shanghai Jiao Tong University
School of Medcine
Shanghai 200025, China
*Chapter 4.4*

**Xin-Yi Zhang**
Department of Pulmonary Medicine
Shanghai First People's Hospital
Jiao Tong University
Shanghai 200080, China
*Chapter 4.1*

**Zhuo-Ran Zhang**
Department of Microbiology
Dalian Medical University
Dalian 116044, China
*Chapter 3.3*

**Chao Zhao**
Key Laboratory of Medical Molecular Virology
Shanghai Medical College
Institutes of Biomedical Sciences
Fudan University
Shanghai 200032, China
*Chapter 2.1*

**Ping Zhao**
Department of Microbiology
Second Military Medical University
Shanghai 200433, China
*Chapter 5.2*

**Xin Zhao**
Department of Hand Surgery
Huashan Hospital
Fudan University
Shanghai 200031, China
*Chapter 4.6*

**Bai-Hui Zhao**
Shanghai Municipal Center for Disease Control and Prevention
Shanghai 200336, China
*Chapter 4.2*

**Li-Xiong Zheng**
Department of Dermatology
Shenzhen People's Hospital
Second Clinical Medicine College of Jinan University
Shenzhen 518020, China
*Chapter 5.1*

**Ya-Jie Zheng**
Department of Dermatology
Shenzhen People's Hospital
Second Clinical Medicine College of Jinan University
Shenzhen 518020, China
*Chapter 5.1*

**Ying-Jie Zheng**
Key Laboratory of Public Health Safety
Ministry of Education
Department of Epidemiology
School of Public Health
Fudan University
Shanghai 200032, China
*Chapter 2.5*

**Wen-Jun Zhong**
Chinese Center for Disease Control and Prevention
Beijing 100050, China
*Chapter 1.2*

**Xin Zhou**
Department of Pulmonary Medicine
Shanghai First People's Hospital
Jiao Tong University
Shanghai 200080, China
*Chapter 4.1*

**Duan-Hua Zhou**
Guangzhou Center for Disease Control and Prevention
Guangzhou 510080, China
*Chapter 3.6*

**Shi-Hang Zhou**
Dalian Red Cross Blood Center
Dalian 116001, China
*Chapter 3.3*

**Xiu-Zhen Zhou**
Guangzhou Center for Disease Control and Prevention
Guangzhou 510080, China
*Chapter 3.6*

**Zhi-Tong Zhou**
Shanghai Public Health Clinical Center affiliated to Fudan University
Shanghai 201508, China
*Chapter 3.1*

**Yue Zhu**
Shanghai Center for Bioinformation Technology
Shanghai 200235, China
*Chapter 2.2*

**Jian-Fu Zhu**
Deqing County Center for Disease Control and Prevention
Deqing 313200, China
*Chapter 2.5*

**Qi-Rong Zhu**
Children's Hospital of Fudan University
Shanghai 201102, China
*Chapter 4.5*

# Index